THE SKY AT NIGHT

TIM B. HUNTER

The SKY at Night

Easy Enjoyment from Your Backyard

THE UNIVERSITY OF
ARIZONA PRESS
TUCSON

The University of Arizona Press
www.uapress.arizona.edu

We respectfully acknowledge the University of Arizona is on the land and territories of Indigenous peoples. Today, Arizona is home to twenty-two federally recognized tribes, with Tucson being home to the O'odham and the Yaqui. Committed to diversity and inclusion, the University strives to build sustainable relationships with sovereign Native Nations and Indigenous communities through education offerings, partnerships, and community service.

ISBN-13: 978-0-8165-4812-5 (paperback)
ISBN-13: 978-0-8165-4813-2 (ebook)

Cover design by Leigh McDonald
Cover photo by tengyart/Unsplash
Designed and typeset by Leigh McDonald in Adobe Caslon Pro 10.5/14,
Millimetre by Jérémy Landes (Velvetyne Type Foundry) and Interstate (display)

Unless otherwise noted, all images are the author's.

Content of this book is based on Tim Hunter's Sky Spy columns, published in the *Arizona Daily Star* beginning in 2007.

Library of Congress Cataloging-in-Publication Data
Names: Hunter, Tim B., author.
Title: The sky at night : easy enjoyment from your backyard / Tim B. Hunter.
Description: Tucson : University of Arizona Press, 2023. | Includes bibliographical references and index.
Identifiers: LCCN 2022015966 (print) | LCCN 2022015967 (ebook) | ISBN 9780816548125 (paperback) | ISBN 9780816548132 (ebook)
Subjects: LCSH: Amateur astronomy. | Astronomy—Popular works. | Astronomy—Observers' manuals.
Classification: LCC QB44.3 .H86 2023 (print) | LCC QB44.3 (ebook) | DDC 520—dc23/eng20220901
LC record available at https://lccn.loc.gov/2022015966
LC ebook record available at https://lccn.loc.gov/2022015967

Printed in the United States of America
♾ This paper meets the requirements of ANSI/NISO Z39.48-1992 (Permanence of Paper).

This book is dedicated to my dear wife, Carol, who allows me to let my hobby of amateur astronomy run amok. It is also dedicated to my grandchildren Braeden Chhatpar and Amelia Chhatpar. May they come to love the sky as much as I do and remember me fondly someday when they are looking at Orion the Hunter, Scorpius the Scorpion, or a crescent Moon nearby ever-brilliant Venus.

Contents

Preface   *ix*

1. Amateur Astronomy   3
2. The Moon   14
3. The Planets   35
4. Stars   51
5. Constellations   84
6. Other Wonders of the Night   110
7. The Seasons and the Calendar   123
8. Selected Famous Astronomers, Events, and Places   138
9. What I Learned from Writing an Astronomy Column   147
10. What Telescope Should You Buy?   162

Acknowledgments   *167*
Astronomical Resources   *169*
Glossary   *171*
Selected Bibliography   *183*
Index   *185*

Preface

I **HAVE BEEN** an amateur astronomer since 1950 when Miss Wilmore, my first-grade teacher, showed me a book of the constellations. I was fascinated by a drawing of Cygnus the Swan and wondered whether I could ever see that in the sky. As the years passed, my interest in astronomy grew. I got a Criterion four-inch Dynascope reflecting telescope when I was in eighth grade. I had saved up $50 from mowing lawns. My father helped me set it up, and I used it for many years off and on through high school, college, and medical school. In 1970, while I was serving in Vietnam, my parents gave my Dynascope to a boy down the street thinking I no longer wanted it or would use it. I often wonder what happened to that telescope and the boy.

My astronomy interest was always there, though I did not want to become a professional astronomer. I wanted to become a physician like my paternal grandfather, who was a pathologist. I graduated from Northwestern Medical School in 1968 and eventually became a radiologist after finishing my radiology training in 1974 at the University of Michigan in Ann Arbor.

Through all of this my interest in astronomy never went away, though it was somewhat on hold and only blossomed again when I moved to Tucson, Arizona, in January 1975 to become a faculty member of the Department of Radiology at the University of Arizona. Tucson likes to claim it is the Astronomy Capital of the World, and I agree. I joined the Tucson Amateur Astronomy Association, Inc. (TAAA) and became very active observing the sky, trying astrophotography, and attending local and national astronomical meetings and events.

Even though medicine was a wonderful career, I wondered whether I should have tried to become a professional astronomer or maybe even an astronaut. In the summer of 1976, I met Frank, a graduate student completing a PhD in engineering. He had a master's degree in astronomy from the University of Chicago, and I asked him why he switched to engineering from astronomy. Frank told me there were few jobs in professional astronomy, and he realized it was best to enjoy astronomy as an avocation rather than as a vocation. This got me thinking. I realized I was much better off enjoying astronomy as a hobby rather than as a vocation. I would have been no good as a professional astronomer, certainly not good enough to win a lot of grants or observing time on large professional telescopes. I had the best of all worlds: a great, satisfying career, and an ever-challenging and enjoyable hobby.

In January 2007, I received a call from Inger Sandal, editor of *Caliente*, the Thursday insert in the *Arizona Daily Star* that tells the public what to do about town, where to dine, what shows to see, and what adventures out of town to anticipate. She wanted to know if I would be interested in writing the Sky Spy column for each weekly edition of *Caliente*. This column had been started more than twenty years previously by Michael Smith, MD, a highly respected neurologist and amateur astronomer. I had known Mike for many years through our membership in the TAAA, as well as from his training at University Medical Center, the academic hospital for the College of Medicine at the University of Arizona.

Mike started Sky Spy on his own initiative, writing to the editor of the *Arizona Daily Star* and suggesting an astronomy column for the public. I don't know if he came up with the name Sky Spy, but he wrote a splendid column for nineteen years. I read it faithfully every week when it came out. After Mike, the column was continued by one of the paper's reporters, who did a very fine job with it for seven years until he left the *Arizona Daily Star* to become the food editor of the rival newspaper, the *Tucson Citizen*.

Inger Sandal, the editor of *Caliente*, was in immediate need of a writer for the column when she called me. I certainly was not on anyone's horizon as a noted astronomy columnist. Inger first asked David Levy, the popular author and comet discoverer, if he would like to write the column. He declined because he had his own column Star Trails in *Sky and Telescope* magazine, as well as many other ongoing writing projects and commitments. David is a very dear friend of mine and he suggested Inger call me.

I was certainly intrigued by Inger's offer, but I didn't know if I was up to the job. She kindly provided me with several recent Sky Spy columns to give me a perspective on what they contained and what she wanted. Then, I worked furiously that afternoon, stealing a bit of university time for my outside interest, writing a couple of sample columns that I sent to her. She liked them and suggested we proceed. On February 1, 2007, my first Sky Spy column was published.

Initially, the column was three hundred words and contained an illustration that I "drew" digitally using Photoshop and various digital star atlases. My illustrations were designed to show what the observer would see when looking in one direction or another. Sometimes, I submitted a photograph with the column. I fancy myself an astrophotographer, and a few of my astrophotographs were wide-field night sky images that would complement a particular column. Fortunately, Inger accepted almost all of my photographs, so I had a good place to publish some of my images that otherwise would not have found a home for publication.

Over time, illustrations were no longer used with the columns except on those rare occasions when I submitted an astrophotograph that I thought was particularly appropriate for a column. Some years after I started, Inger asked me if I could write the columns twice a month rather than weekly, as the newspaper wanted to save a bit of column print space. Instead, I proposed that it continue weekly, but reduced to 250 words so as not to lose reader interest. This worked well, and up to the present time (2022) the column has been published weekly with approximately 250 words per column.

I have now over 750 columns, and this book is a compilation of them. They have not just been thrown all together and published as a book. I have picked and chosen the important points from the columns and collated them, I hope, into an intelligible whole to be enjoyed by the reader. Terms that appear in the glossary are in **bold** when they first appear, while other key terms are in *italics* when they first appear.

The sky is wonderful. It is to be enjoyed day and night, with easy viewing of the night sky the focus of this book. I assume you, the reader, are literate and interested in the sky, but are not particularly knowledgeable. Emphasis is on naked-eye viewing, with an occasional reference to using binoculars or a small, low-power telescope. I assume you are not familiar with most of the constellations, but I hope the descriptions and the directions I provide are good enough to help you find your way around the sky. I will have succeeded if you enjoy the sky as much as I do and make friends with the Moon, planets, and stars.

THE SKY AT NIGHT

1

Amateur Astronomy

CALL EVERYONE who reads this book an *amateur astronomer*, a lover of the night sky. You can take the sky in as much or as little as you want. Some amateur astronomers have a PhD in astronomy and related sciences and do astronomy professionally for money. Many others like to walk outside at night occasionally to see something interesting, but otherwise are not bothered by astronomical thoughts. Either way is fine with me.

I like the Moon for its beauty and intrigue. I also like to use it as a pointer to other celestial objects. Everyone can find the Moon. If something is close to it, the Moon is an excellent guide to that object.

In this book, all of the material is applicable to almost any observing location in the Northern Hemisphere. Many of the observing tips, astronomical definitions, and other descriptions apply to both the Northern and Southern Hemispheres, but the **constellations** and other phenomena described herein are best seen in the Northern Hemisphere. Date-specific or location-specific phenomena, such as solar or **lunar eclipses**, are not mentioned except in general terms as part of a wider discussion of the sky.

Any time of day or night is fair game for an observing session, though there is an emphasis on early evening viewing, often in the evening **twilight**, as that is more convenient for most people. However, one should not shy away from the predawn sky, and sometimes it is necessary to be up in the middle of the night to see something interesting or unusual.

Sleep deprivation is definitely a worry for an amateur astronomer. Some of the best astronomical events happen in the middle of the night on a work night or a school night. You either can lose sleep or miss the event. While amateur astronomy cannot be considered a hazardous hobby, it is not without risks. If one is tired and sleep deprived, it is easy to fall off a ladder while looking through a telescope or to walk into sharp objects in the dark.

Driving home in the morning after being up all night observing can be very dangerous not only to the astronomer but also to his or her passengers and other motorists. You must use your common sense and make sure you do not attempt to work, operate dangerous machinery, or drive while sleep deprived. In addition, alcohol and smoking are not good for observing.

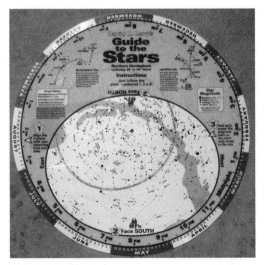

FIGURE 1.1 Planisphere.

OBSERVING AIDS

How does one learn the constellations, **planets**, and important bright **stars**? With patience and practice. Observing aids are helpful. The best observing aid by far is a knowledgeable friend or family member who shows you the sky.

PLANISPHERE

Beyond a helpful family member or friend, by far the most useful aid for learning the night sky is a **planisphere** (figure 1.1).

This is a cardboard or plastic chart showing the constellations. The chart rotates inside of a square or rectangular border that is labeled with the time and date. To use a planisphere, you take it outside and hold it up overhead and orient it north–south and east–west according to its directions. Dial in the present time and date. The planisphere chart shows you the constellations currently visible. By looking back and forth between the planisphere and the sky, you can rapidly recognize constellations and prominent stars. Planispheres cost $10–$30 and may be purchased from many bookstores and online vendors. They are invaluable. I use one all of the time.

MOBILE APPS

What about mobile apps that show the sky? They are great, and I have several of them on my phone. However, I don't find them as useful as a planisphere. Even the biggest cell phone or tablet is nowhere near as large as a typical planisphere and does not give the overview and field of view of a planisphere. However, mobile apps are great for identifying a specific bright object, such as a planet, as planispheres are not useful for finding them. Carefully point the cell phone at the object in question, and a good app will identify what it is. You must be most careful with your aim. Mobile apps are up-to-date for time and location and are superb for identifying the planets as well as bright stars.

Use whatever works best for you, a planisphere or a mobile app. Or better yet, both.

BINOCULARS

Binoculars are wonderful for sky viewing. What is even better is to have a tripod to steady the binoculars if they are above 8–10 power or are heavy. I won't recommend a specific brand other than to say you should choose a pair that is easy to hold and works well during the day for watching a football game or for bird-watching. The binocular tripod is most convenient when you want to study an object, such as the moons of Jupiter or the large Beehive **star cluster**, beyond just glancing at it.

I would not exceed 7 or 8 power binoculars for most observing. Higher-power binoculars or very large binoculars with objective lenses larger than 50 mm in diameter are quite fine for observing fainter celestial objects. They cannot be held for long by hand due to their weight, and their higher power magnifies any vibration or unsteadiness in holding them, thereby ruining most observing sessions. They require a good heavy-duty tripod for steadiness and ease of use and are more expensive.

TELESCOPES

The Sky Spy columns and this book are predicated on the notion that you do not own a telescope or have access to one. There is an occasional reference to viewing something with a small, low-power telescope, but one can spend a lifetime enjoying the sky with only naked-eye observing.

A common mistake is to buy a cheap "dime-store" telescope that doesn't work well or, even worse, to spend a considerable amount of money on a useful instrument without learning how to use it or be unwilling to invest the considerable time necessary to use the telescope to its best advantage.

Personally, one of the saddest objects I can imagine is a good telescope that is never used, tucked away collecting dust in a closet or attic. Such forlorn instruments are a waste of someone's hard-earned funds or a cause of great disappointment. Avoid this mistake. There is a later, more detailed discussion of what telescope, if any, you should buy in chapter 10.

If you really want to buy a telescope and know what you are doing, great. Go ahead and make that purchase. Otherwise, put off buying a telescope, or skip ahead and read my more detailed discussion in that regard. Meanwhile, look for a local amateur or amateur astronomy club that has telescope viewing from time to time for the public. You can look through fine telescopes for hours for free, enjoying any number of celestial delights along the way. What is more, you can learn about telescopes, which would help you decide if owning one is right for you.

IMPORTANT OBSERVING TERMS

There are some often used astronomical terms that are very important to learn. Fortunately, there are not that many of them, and they are easy to learn. In fact, most people are probably already familiar with most of them. Some of them are used in everyday life, and others are relevant only for astronomical discussions. Occasionally, the use of a term in everyday life differs somewhat from how it is used astronomically.

ANGULAR DISTANCE

The distance from one horizon to the opposite horizon is 180 degrees. An object overhead is 90 degrees from the horizon in all directions. When you extend your hand to arm's length, your index finger is about 1 degree in width, and the distance across your palm is 10 degrees in width when projected against the sky.

APOGEE AND PERIGEE

The terms **apogee** and **perigee** refer to any elliptical orbit. They describe when an orbiting body is farthest from (apogee) and closest to (perigee) the object it is orbiting. Thus, there is a monthly apogee and perigee for the Moon orbiting Earth and a yearly apogee and perigee for Earth orbiting the Sun.

AZIMUTH AND ALTITUDE

Azimuth and **altitude** are part of the *horizontal coordinate system.*

Altitude is a measure of the angular distance of an object above or below the horizon. Objects located on the horizon have 0 degrees altitude and, at the zenith (directly overhead), 90 degrees altitude. In formal astronomical circles, objects below the horizon have negative altitudes.

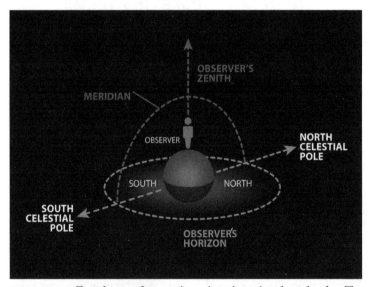

FIGURE 1.2 Zenith, meridian, and north and south **celestial poles.** The north and south celestial poles are projections of Earth's poles into the night sky. Drawing by Debra Bowles.

Azimuth is based on the 360 degrees of a traditional compass with due north at 0 degrees, due east at 90 degrees, due south at 180 degrees, and due west at 270 degrees.

If a celestial body has an azimuth of 180 degrees and an altitude of 45 degrees, it is directly south and is halfway between the horizon and directly overhead (zenith).

CELESTIAL COORDINATE SYSTEM

Astronomers use a **celestial coordinate system** to define the position of an object in the sky. It is an equatorial system, in which **right ascension** is the equivalent of longitude and **declination** is the equivalent of latitude. This system is more confusing and uses somewhat different units than the familiar longitude and latitude system. More commonly, a column such as Sky Spy uses a horizontal coordinate system to define an object's place in the sky.

CELESTIAL EQUATOR

The **celestial equator** is the projection of Earth's equator into the sky. It divides the sky into the northern and southern celestial hemispheres.

CONSTELLATION

Constellations are just groupings of stars used to identify a region of the sky. The name of each constellation is derived from a mythical being, an object, or a real animal. Some constellations predate written records, and most originally came from the myths and legends of Mesopotamia, Babylon, Egypt, and Greece. A few of the constellations were devised by astronomers in as late as 1750. In 1930, the **International Astronomical Union (IAU)** specially defined a set of eighty-eight constellations and their boundaries in the sky. This official definition of our "modern" constellations is used by professional and amateur astronomers today.

CRESCENT

Crescent describes the curved shape of the Moon or another planetary body that is less than half lit.

ECLIPTIC

Because Earth revolves around the Sun in a year's time, the Sun appears to travel through the sky on a path known as the **ecliptic**. Most of the planets appear close to the ecliptic, and their paths around the sky are confined to the narrow zone known as the **zodiac**, which stretches to about 9 degrees on either side of the ecliptic.

ELLIPTICAL ORBIT

The orbit of an astronomical body, such as the orbit of the Moon revolving around Earth, is the path the object follows in its motion around another body. All astronomical orbits are technically ellipses, though they are often very hard to distinguish from a circle. In fact, a circle is a special kind of ellipse.

ELONGATION

Elongation is the angular separation (distance) between the Sun and a solar system body. It is measured in degrees eastward or westward of the Sun. If Venus, for example, has a 10-degree eastern elongation, it is 10 degrees from the Sun and is visible in evening twilight. If it has a 10-degree western elongation from the Sun, it is visible in the morning predawn twilight.

EQUINOX

At the exact time of an **equinox** in the Northern Hemisphere, the Sun crosses the celestial equator from the northern sky to the southern sky (autumnal, September, or fall equinox) or from the southern sky to the northern sky (vernal, March, or spring equinox). Equinox means "equal night"—equal hours of day and night. In fact, the hours of day and night are close to, but not exactly, equal at the time of the equinoxes.

GIBBOUS

Gibbous refers to the **phase of the Moon** or another planetary body that is more than half lit but not completely lit as the **full Moon**.

MERIDIAN

The sky is like an inverted bowl with an imaginary line, the meridian, running from the north to the south celestial pole through the observer's zenith. See figure 1.2. The celestial poles are the imaginary projection of Earth's poles into the sky. Therefore, the meridian divides the sky into eastern and western halves, and when a star or planet crosses the meridian (*culminates*), it is equidistant from the eastern and western horizons.

SOLSTICE

The yearly movement of Earth around the Sun makes the Sun appear to slowly change its place in the sky from day to day if you look for the Sun at the same time every day. The **solstices** are those points in the Northern Hemisphere at which the Sun is at its greatest distance north (summer solstice) or south (winter solstice). Solstice means a culminating point or turning point. The solstices take place at a brief, specific time, but the day upon which a solstice occurs is often also referred to as the solstice or solstice day.

The summer solstice takes place around June 21 every year, and the winter solstice takes place around December 21 every year. The solstices are reversed in the Southern Hemisphere, making the seasons there opposite those of the Northern Hemisphere.

TWILIGHT

After **sunset** there is a period of time in which the sky gradually darkens, and before **sunrise** there is a period of time during which the sky gradually brightens. Twilight, the time between day and night, is caused by scattered sunlight in the atmosphere that illuminates the sky and the ground when the Sun is just below the horizon.

Twilight has strict astronomical definitions. *Civil twilight* is the time when the center of the Sun's disk is less than 6 degrees below the horizon. During this time, the brightest stars and planets appear. At *nautical twilight*, the Sun is between 6 degrees and 12 degrees below the horizon. Prior to modern electronic equipment, sailors used this time for reliable sightings of well-known stars for navigation. *Astronomical twilight* takes place when the center of the Sun is more than 12 degrees and less than 18 degrees below the horizon. The sky appears nearly fully dark, but very faint objects may still be hidden by the dim sunlight in the atmosphere. The length of twilight is very dependent on one's latitude. At high latitudes it can last for hours, but at the equator it can be as short as twenty minutes.

WAXING AND WANING

In this book **waxing** and **waning** usually refer to the lit portion of the Moon. When the Moon's lit portion is increasing, it is said to be waxing, and when the Moon's lit portion is decreasing, it is said to be waning. These terms can also be used generically for other celestial bodies.

ZENITH

Directly overhead. See also the definition for azimuth and altitude.

ZODIAC

The zodiac is a narrow zone in the sky within which the planets appear to travel. It spans about 9 degrees on either side of the ecliptic. The twelve main constellations through which the ecliptic and the Sun and planets pass are the zodiacal constellations: Aries, Taurus, Gemini, Cancer, Leo, Virgo, Libra, Scorpius, Sagittarius, Capricornus, Aquarius, and Pisces. The signs of the zodiac were established several thousand years ago when astrology and astronomy were closely related. However, modern astronomers no longer follow the tenets of astrology, and the two disciplines have gone their separate ways. Formal constellation names and boundaries were standardized by the IAU in 1933 to prevent confusion. In this process, the zodiac gained a thirteenth constellation, Ophiuchus, because its modern boundary slightly overlaps the zodiac.

2

The Moon

THE MOON weaves its way through myths, legends, fables, campfire stories, classic literature, songs, hymns, and poems. It is an important artistic and aesthetic component of every culture, and its movement in the sky is the basis for most early calendar systems. The Moon is an object of beauty and fascination unrivaled by most other objects in nature. It is the only celestial body whose surface features can be seen from Earth with the naked eye, and it is the only other world upon which humans have walked.

The Moon is of great importance to us culturally, astronomically, and probably biologically. The Moon itself has no life. The Moon is a major determinant of our tides, and there is good evidence the Moon is a stabilizing influence on the tilt of Earth's axis. Earth's axis is the imaginary line running through the poles around which Earth rotates. The Moon is relatively large compared to its parent planet, and its tugging on Earth stabilizes Earth. The axis of Earth is presently tilted 23½ degrees with respect to the plane of its orbit. This tilt is responsible for our seasons. It varies by a small amount over a period of twenty-six thousand years, and it has been the same for millions, perhaps billions of years. This keeps most of Earth just right for life—not too hot and not too cold. The axes of many of the other planets are known to have changed drastically in the last few million years. Such changes on Earth would probably doom all forms of advanced life. The next time you look at the Moon, enjoy it for its beauty but also tip your hat to it to thank it for keeping Earth "just right."

The Moon is Earth's "natural" satellite, as opposed to the thousands of objects that have been put into orbit around Earth since Sputnik was launched in 1957. The Moon is nearly as old as Earth, over four billion years in age. Except for the Sun, the Moon is the brightest object in the sky.

The Moon does not emit its own light. It is simply lit by reflected sunlight. If you pass the light of the Sun through a prism, it will be stretched out into a spectrum of colors from deep red to deep purple. This spectrum contains many thin dark lines caused by the various chemical elements in the Sun. The spectrum of the Moon is the same as that of the Sun, confirming that moonlight is the same as sunlight. All of the planets are lit by reflected sunlight. If you take a long exposure of a landscape lit by the Moon, you will get a picture that looks as if it were taken in the daytime.

I love observing the Moon in all of its phases. Some mention of the Moon is found in almost all of my Sky Spy columns. To many astronomers, the Moon is an object of annoyance. Its brightness blots out dimmer objects that are often of more interest. Even so, I think of the Moon as always having something interesting to observe and enjoy, though at times I am glad when it is out of the sky.

It is hard at first to learn the constellations, planets, and bright stars. But everyone can find the Moon, even in its thinnest crescent phase. If a planet, star, or constellation of interest is near the Moon, it is much easier to find than usual, because you can look at the Moon and then shift your gaze in the direction of the object, even though it will be considerably dimmed by the Moon. Try to always enjoy the Moon.

LUNAR MONTH

The Moon moves west to east around Earth, completing one revolution against the **background stars** in about 27.32 days (a *lunar sidereal month*); that is, the Moon comes back to the same approximate position in the starry sky every 27.32 days. The Moon makes one revolution relative to the Sun in about 29.53 days (a *lunar synodic month*). The difference between the two lunar periods (sidereal versus synodic, or starry background versus Sun position) is due to Earth's motion around the Sun. As the Moon is moving around Earth, Earth is moving west to east around the Sun. The Moon does not complete a full lunar synodic cycle until Earth reaches a point in its orbit where the Sun and the Moon are in the same relative positions. Since we are more interested in the Moon's position relative to the Sun, we use the generic term **lunar month**, which is the time between successive **new Moons** and is generally considered to be 29.5 days.

As the Moon revolves around Earth, it is also rotating on its axis at about the same speed it goes around Earth. As a result, the Moon keeps the same face toward Earth. The near side of the Moon is the portion facing Earth, and the far side of the Moon is the portion facing away from Earth. Because the Moon completes one rotation around its axis in 29.5 days, a lunar month, it has a day side and a night side like Earth does, though its "day" is 29.5 Earth days long instead of twenty-four hours like a full day on Earth.

The unlit or night side of the Moon is the dark side of the Moon. This should not be confused with the far side of the Moon, the side of the Moon facing away from Earth. Both the near side of the Moon and the far side of the Moon have equal periods of day and night. When we look at a partially lit Moon, either a crescent Moon or a gibbous Moon, we view a portion of the Moon's daytime side and a portion of its nighttime side.

Due to variations in the Moon's speed around Earth and due to a slight inclination of the Moon's orbital plane with respect to the plane of Earth's orbit, we can peer around the Moon's "edge" a bit from time to time, which allows us to see more than 50 percent of the Moon's surface over many days' time. At any given moment, however, we can only see 50 percent of the Moon's surface from Earth. The 50 percent portion of the Moon's surface that is visible at any given time varies slightly from day to day, allowing us to see up to 59 percent of the total surface of the Moon over an extended period. The combination of effects that allow us to see slightly more than half of the Moon's surface is called *lunar libration*.

PHASES OF THE MOON

Everyone knows the Moon goes through phases and sees these on a nightly basis whenever the Moon is visible. From night to night the Moon never looks the same—the Moon's lit and unlit portions are constantly changing. This is most obvious if you look at the Moon from one night to the next, but on a given night if you follow the Moon closely with binoculars or a small telescope, you can see obvious, albeit subtle changes in the Moon's illumination after only a couple of hours.

While everyone knows about the Moon's phases, few understand what causes them or know their proper names. I have been an active amateur astronomer since first grade. Frankly speaking, I did not understand the Moon's phases until I was in high school. The hundreds of diagrams showing the Moon's phases found in almost any astronomy textbook, or nowadays on the web, just did not do it for me. The Moon is a sphere in space, going around another sphere (Earth) while being lit by a bright distant sphere (the Sun). This three-dimensional reality is poorly illustrated by a simple two-dimensional drawing.

One day while watching television in a darkened room and doing high school homework (a combination of activities that did not produce good results, by the way), I was fiddling with a tennis ball and waving it around for no good reason. Light coming from the television lit a portion of the ball more than the rest, and I suddenly realized how the lighting on the ball resembled the phases of the Moon as I swung the ball around myself. The ball got no light when it was between me and the TV set (equivalent to new Moon), while it was fully lit when it was behind both me and the TV (equivalent to full Moon). When the ball was to my side, its front half was lit, and its back half was not lit by the TV (equivalent to a quarter Moon). To really get a three-dimensional feeling for the phases of the Moon, set up a bright light on one side of a darkened room. Stand on the other side of the room with a tennis ball or a larger ball and swing it around yourself just like I described. You can easily simulate the phases of the Moon in a three-dimensional fashion such that you can understand them much better than if you were to study a diagram in a book.

Figure 2.1 shows the cardinal directions for the sky for the Northern Hemisphere. The east and west sides of the Moon are indicated in black text. The Moon moves continually from west to east. This motion is evident from day to day, as the Moon's position in the sky is different from day to day, and its phase changes.

Standard astronomical maps and photographs like figure 2.1 have north at the top and west to the right. This is how the sky map would look if you took it outside and held it up toward the sky with your back to the north. North would be at the top of the map and toward your head, east would be toward your left, west toward your right, and south toward your feet at the bottom of the map.

Whenever you use an illustration to find your way in the sky, it is most important to make sure you fully understand the direction labels on the illustration. Astronomical maps from earlier times until the middle of the twentieth century frequently had south at the top, because when viewed through a telescope, objects often appeared upside down. Obviously, this can get confusing very quickly if you do not understand the cardinal directions (north, south, east, and west) shown on an astronomical map or if the orientation of the map is not clearly marked.

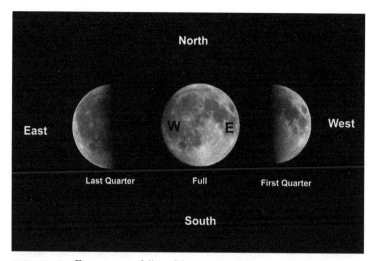

FIGURE 2.1 First quarter, full, and last quarter Moons with sky directions (white) and Moon directions (black) shown.

NEW MOON

Day 0 of a lunar month is new Moon. New Moon is when the Moon is directly between Earth and the Sun, and we see the dark (unlit) side of the Moon. On rare occasions when the Moon is exactly lined up between the Sun and Earth, an eclipse of the Sun is visible from part of Earth. Most times the Moon is slightly above or below the plane of Earth's orbit, so no eclipse occurs, but the portion of the Moon facing Earth is dark.

FIRST QUARTER, LAST QUARTER, AND FULL MOON

Days 1–6 of a lunar month represent a waxing crescent Moon. Waxing means increasing in size. A waxing crescent Moon is a Moon whose lit portion is crescent shaped, with the crescent growing each day. Seeing a very young (one day old or less), thin crescent Moon after new Moon is not only a fun challenge, but also important for various lunar calendars (figure 2.2). The Hebrew calendar and the

FIGURE 2.2 Very young Moon at 17 hours 58 minutes. You can see the faint crescent just above the telephone wires to the right of center.

Islamic calendar are based on the first sighting of a very thin lunar crescent after new Moon.

Once the lit portion of the Moon has the configuration of a semicircle, it is a first quarter Moon (figure 2.1). This occurs between days 7 and 8 of a lunar month. As the lit portion of the Moon grows even larger, it becomes a waxing gibbous (more than half lit) Moon until it reaches a full Moon, which occurs between days 14 and 15. Then, as the lit portion of the Moon begins to shrink, it is a waning (shrinking) gibbous Moon. Once the lit portion again reaches a semicircle configuration between days 21 and 22 of a lunar month, it is a last quarter Moon. A last quarter Moon is also sometimes referred to as the third quarter Moon. As the lit portion of the Moon shrinks further, the Moon becomes a waning crescent until it disappears, becoming a new Moon. Then, it becomes a waxing crescent Moon to start the cycle all over again. The time from one new Moon to another new Moon is approximately 29½ days, a lunar month, as mentioned previously.

A first quarter Moon technically means the Moon is one quarter of the way around Earth from one new Moon to the next new Moon. Last quarter or third quarter Moon technically means the Moon is three quarters of the way from one new Moon to the next new Moon.

A waxing crescent Moon is evident in the evening sky after sunset. From our perspective, the right side (the east side) of the Moon is lit. By first quarter, the Moon is approximately on the center line of the sky (the meridian) at sunset around 6:00 p.m. local time, and it sets at around **midnight** local time. In this context, local time means the solar time at your location regardless of the time zone.

A full Moon is 180 degrees away from the Sun, on the opposite side of Earth from the Sun. Therefore, it rises in the east as the Sun sets in the west. It crosses the meridian at around midnight local time. The full Moon sets in the west as the Sun rises in the east.

A last quarter Moon has its left side (its west side) lit from our perspective. It rises at approximately midnight local time and sets at approximately **noon** local time. At noon local time, the Sun is directly to the south and as high above the horizon as it gets.

This discussion about crescent and gibbous Moons, first quarter Moon, last quarter Moon, and full Moon can be quite confusing and overwhelming at first. Don't despair. Simply take the time to look at the Moon most evenings after sunset or mornings before sunrise. You will soon recognize the patterns just described, particularly if you occasionally look back at this discussion or a similar one. After a while, you will be able to recognize the phase of the Moon at a glance, knowing its approximate age in days by its appearance and by the hour you are observing it.

SUNRISE, SUNSET, MOONRISE, AND MOONSET

Sunrise, sunset, moonrise, and moonset are words learned at an early age and used in everyday conversation. However, it is unlikely most adults are truly familiar with the precise definitions for these terms. The United States Naval Observatory defines sunrise and sunset as follows (https://aa.usno.navy.mil/faq/RST_defs): "Sunrise and sunset conventionally refer to the times when the upper edge of the disk of the Sun is on the horizon[, considered unobstructed relative to the location of interest] . Atmospheric conditions are assumed to be average, and the location is in a level region on the Earth's surface." Moonrise and moonset are defined in the same manner for the upper edge of the disk of the Moon.

How long do sunrise and sunset take, and how long do moonrise and moonset take? They are all brief moments in time, but one often considers them to be the time it takes the respective disk of the Sun or Moon to clear the horizon. In other words, how long does it take for the rising Sun's disk to completely appear above the eastern horizon or the setting Sun's disk to completely disappear below the western horizon? Similarly, how long does it take for the rising Moon's disk to completely appear above the eastern horizon or the setting Moon's disk to completely disappear below the western horizon?

In the case of the rising or setting Sun, the two main factors that determine "the length of sunrise" or the "length of sunset" are one's latitude and the angle of the rising or setting Sun with respect to the horizon. Sunrise and sunset are faster near the equator than in more northern or southern latitudes. For most populated regions in the world at midlatitudes, sunrise and sunset take between two and five minutes. This is the time at sunrise from when the upper edge of the disk of the Sun is on the eastern horizon to when the lower edge of the Sun's disk clears the eastern horizon. At sunset this is the time from when the lower edge of the Sun's disk touches the western horizon until the upper edge of its disk is at the western horizon.

The issue of sunrise and sunset quickly gets complicated when you are at or above the Arctic or Antarctic Circles. At these locations, there are periods in summer when the Sun is continually above the horizon (the "midnight Sun") and periods in winter when the Sun never rises.

How long it takes for the Moon to rise or set leads to a similarly complex discussion that must take account of one's latitude and the season. The Moon continually moves from west to east on its daily trek through the sky as it revolves around Earth in 29½ days. However, the Moon's apparent motion is due more to Earth's rotation than to the Moon's actual movement against the background sky. On average, it takes the Moon about two minutes to move its width through the sky, meaning "moonrise" and "moonset" last that long. This is a fairly rapid motion, and it is why you can detect the Moon's movement if you watch it rising or setting with well-mounted binoculars or a low-power telescope.

MOON ILLUSION

The full Moon or a nearly full Moon looks especially prominent as it rises above the horizon. It looks larger than when it is higher in the sky. This is a poorly understood optical illusion. Objective measurements of the Moon's apparent diameter at the horizon and when it is overhead show no significant differences. Why the Moon appears larger near the horizon has been the subject of multiple studies and theories, but there is no generally accepted explanation for this effect.

LUNAR OCCULTATIONS

As the Moon moves through the sky from west to east, it occults (covers) stars along its path. This happens continuously, but it is not observed very often, even by those who spend a lot of time observing the Moon through a telescope. Most of the occulted stars are faint, and often they are obscured from view by the relative brightness of the Moon.

Occasionally, the Moon occults a bright star along its path, such as Aldebaran in Taurus the Bull or Regulus in Leo the Lion. The Moon even occasionally occults a planet. Viewing an **occultation** through a small telescope or high-powered binoculars on a steady tripod is fun and, in some cases, can even contribute to scientific knowledge. An excellent source of information about lunar occultations and other types of occultations is the International Occultation Timing Association (https://occultations.org/).

METONIC CYCLE

The last time the Moon was full on New Year's Day occurred in 2018. Before that, it occurred in 1999. The next time it will happen is 2037. It turns out that every nineteen years the full Moon occurs on January 1, and a second full Moon (sometimes known as a "**blue Moon**") occurs on January 31.

You can choose any date of the year, and the phase of the Moon on that date repeats every nineteen years. The Hebrew calendar, which is based on the Moon, observes that the Sun, Earth, and Moon come back into the same relative positions every nineteen years. This relationship is known as the Metonic cycle. Meton of Athens introduced the concept in 432 BC, though probably it was known by Neolithic peoples more than two thousand years earlier. This cycle represents the complex relationship between Earth's revolution around the Sun every 365¼ days and the length of the lunar month of 29½ days. Moreover, these relationships are superimposed on our modern Gregorian calendar, which defines a year as 365 days plus a leap year day every four years on even years.

We have modern technology able to measure time in microseconds and predict celestial events quite accurately. Nonetheless, the ancients were just as clever as we are and much better observers of the sky.

FULL MOON NAMES

There are many, many names for the full Moons of the year. According to the *Farmers' Almanac*, Native Americans of the northern and eastern part of North America kept track of the seasons by naming the recurring full Moons and their months: January–Wolf Moon; February–Snow Moon; March–Worm Moon; April–Pink Moon; May–Flower Moon; June–Strawberry Moon; July–Buck Moon; August–Sturgeon Moon; September–**Harvest Moon**; October–**Hunter's Moon**; November–Beaver Moon; December–Cold Moon, or Long Nights Moon. Because the lunar month is only 29½ days long, the full Moon dates constantly shift from year to year.

Different Native American tribes have different names for the full Moon. Moreover, colonial settlers in America had a list of full Moon names, such as Planter's Moon (April) or Christmas Moon (December). The Chinese also had many interesting names for full Moons, such as Peony Moon (April), Lotus Moon (June), Chrysanthemum Moon (September), and Bitter Moon (December). The list of names for full Moons is almost endless because most cultures both ancient and modern have given the full Moon many names. A somewhat more modern example is the blue Moon, which is the third of four full Moons in a season or the second full Moon in a month (see below).

HARVEST MOON AND HUNTER'S MOON

A Harvest Moon is the full Moon nearest to the autumnal equinox. In general, the Moon rises about fifty minutes later each day, but at around the time of the autumnal equinox, the Moon rises approximately thirty minutes later from one night to the next. The Moon seems to hang along the eastern horizon from night to night after sunset, providing extra light for "the last days of summer."

The Hunter's Moon is the first full Moon after the Harvest Moon and, like it, seems to hang above the eastern horizon longer than usual after rising. The extra light provides more time for hunters to look for their prey and more time for farmers to bring in crops. In the Northern Hemisphere the Harvest Moon usually occurs in September, but it can occur in October.

While the Harvest Moon seems to be larger, brighter, and more yellow than other full Moons, this is an illusion. The Moon is no larger or brighter than usual, but we may perceive it as more yellow or orange, because we tend to notice a Harvest or Hunter's Moon as it rises over the mountains or hills. It is low in the sky, and its blue light is scattered more than its red light, which gives it an orange or yellow color.

BLUE MOON

A blue Moon is a common term for a rare event, as in "once in a blue Moon." However, astronomically speaking, blue Moons are not that rare, depending on how you define the term. Most years have twelve full Moons, but approximately every three years there is a thirteenth full Moon. The extra full Moon is called a blue Moon. It is not necessarily the last full Moon in the year, because as we noted each full Moon in many cultures was given a folklore name according to its time of the year, such as the Hunter's Moon. An extra Full Moon that did not fit into the scheme for one of these folklore names was called a blue Moon.

Most seasons (spring, summer, fall, and winter) have three full Moons. Occasionally, there will be four full Moons in a season. The third full Moon in this case is sometimes called a blue Moon. The most common definition for a blue Moon is a second full Moon occurring in a calendar month. Another, looser definition for a blue Moon is a Moon that appears blue due to unusual atmospheric conditions.

SUPERMOON

A **supermoon** occurs if the Moon is at its closest approach to Earth when it is full. The average distance between the center of Earth and the center of the Moon is 239,228 miles (385,001 kilometers). This distance varies from a low of 221,500 miles (356,500 kilometers) at perigee (when the Moon is closest to Earth) to a high of 252,700 miles (406,700 kilometers) at apogee (when the Moon is farthest from Earth).

A supermoon is not all that super. The Moon may appear to be a bit larger and brighter. A full Moon at perigee can be up to 14

FIGURE 2.3 (a) Supermoon of November 13, 2016, compared with the full Moon of March 23, 2016, when the Moon was near apogee. (b) Full Moons from November 13, 2016, to December 2, 2017, with the distance from Earth in kilometers at the time the image was taken. The October 6, 2017, Moon was one day past full.

percent larger in diameter than a full Moon at apogee. Of course, most of the time the full Moon is somewhere in between perigee and apogee, and the difference in appearance from one full Moon to the next is slight. Figure 2.3a shows a supermoon compared with a full Moon at apogee. Figure 2.3b shows a year's worth of full Moons at the same image scale so you can see how the Moon's apparent size and its libration changes from month to month. A full Moon at apogee is sometimes called a *micromoon*. The supermoon on November 13, 2016, was at a distance of 221,526 miles (356,511 kilometers). It was the closest supermoon since 1948. While the supermoon phenomenon is interesting, it is of no astronomical importance. It does encourage us, however, to appreciate our Moon, which is always gorgeous and far more complex than it first seems.

LUNAR FEATURES

CRATERS AND MARIA

With the naked eye and low-power binoculars it is possible to see many features on the Moon, particularly two prominent craters, Copernicus and Tycho (figure 2.4). Tycho is especially bright and interesting because its rays can be seen streaking over a large portion of the Moon. The rays represent material spewed out of Tycho during the incredible impact that produced this crater approximately one hundred million years ago. It is a very "fresh" crater, and the material from its creation lies over the older lunar surface. Copernicus is a gorgeous crater best seen through a telescope, but it is visible through binoculars.

There are many large dark patches on the Moon called maria (singular mare). The word means "sea" in Latin, as the first astronomers to view the Moon through telescopes in the 1600s thought the maria were giant bodies of water. Now we know they are large expanses of dark lava from ancient eruptions on the Moon. They are easily recognizable to the naked eye, and they have very fanciful names, such as Mare Crisium, the "Sea of Crises," and Mare Imbrium, the "Sea of Cold." Every time I look at the Moon, even when it is not full, I try to identify the various maria. They are hard to remember, but they are fun to observe.

FIGURE 2.4 Full Moon with major features labeled. Craters (black): Copernicus [C]; Tycho [T]. Maria (white): Mare Crisium [MC], "Sea of Crises"; Mare Fecunditatis [MF], "Sea of Fertility"; Mare Nectaris [MN], "Sea of Nectar"; Mare Tranquillitatis [MT], "Sea of Tranquility," where Apollo 11 landed; Mare Serenitatis [MS], "Sea of Serenity"; Mare Frigoris [MF], "Sea of Cold"; Mare Imbrium [MI], "Sea of Rains"; Mare Humorum [MH], "Sea of Moisture"; Oceanus Procellarum [OP], "Ocean of Storms"; Mare Nubium [MNu], "Sea of Clouds."

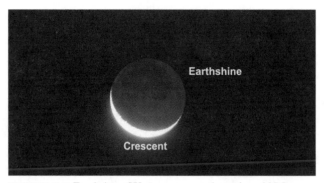

FIGURE 2.5 Earthshine. Waxing crescent three-day-old Moon.

EARTHSHINE

When the Moon is in a thin crescent phase, there is a visible glow from the dark portion of the Moon. This is **earthshine**—light from the Sun that has reflected off of Earth onto the Moon and back to Earth (figure 2.5). Earthshine is also sometimes called the "Moon's ashen glow." Earthshine is usually visible a few days before new Moon (roughly days 24–27), when the Moon is a waning crescent, and a few days after new Moon (roughly days 1–4), when the Moon is a waxing crescent on the way to first quarter.

The lit crescent part of the Moon is the daylight side of the Moon as seen from Earth, and the dark portion of the Moon is its night side. Leonardo da Vinci is credited with explaining earthshine in the early 1500s when he realized sunlight is reflected off of both Earth and the Moon.

The earthshine visible with a waxing crescent is sometimes referred to as the "old Moon in the new Moon's arms." The old Moon is the gibbous unlit portion of the Moon, and the new Moon is the bright waxing crescent. The earthshine visible with a waning crescent Moon is sometimes called "the new Moon in the old Moon's arms." In this case the waxing gibbous unlit portion of the Moon is growing, and the waning lit crescent portion of the Moon is what is left of the old Moon. These terms are confusing and often used incorrectly, even in popular books and on websites. It is best not to worry about them. Just enjoy the earthshine associated with a crescent Moon.

TERMINATOR

The first quarter Moon is bright, and earthshine is no longer apparent on the dark portion of the Moon. The Moon at this phase is wonderful through binoculars, but it is a must-see through a small telescope. No matter the phase of the Moon, there is a "line," the **terminator**, which divides the lit (day side) from the unlit (night side) of the Moon. If you could stand on the Moon at the terminator, you would see the Sun very slowly rise on the Moon's eastern horizon.

Terminator is an old astronomical term and is not to be confused with Arnold Schwarzenegger and the *Terminator* movies. The mountains and craters along the terminator cast long shadows because the Sun is very low in their sky. Through a telescope this region is quite dramatic: the craters stand out with startling contrast and very long shadows. When the Moon is full, there is little shadowing, and it is difficult to see most craters.

MOON HALOS

Sometimes, on rainy, misty evenings a ring or halo can be seen around the Moon (figure 2.6). It is a wonderful sight and not that uncommon. I probably see a Moon halo several times a year, even in hot dry Tucson. A similar halo can be seen around the Sun on an occasional cloudy day (figure 2.7).

These halos occur when high thin clouds that contain tiny ice crystals cover much of the sky. Each of these crystals acts like a tiny lens, and many of the crystals have a similar elongated hexagonal shape. As a result, each crystal bends the Moon's light in a similar direction, producing a 22-degree ring of light around the Moon. If you look carefully, you will note that the inner edge of the halo has

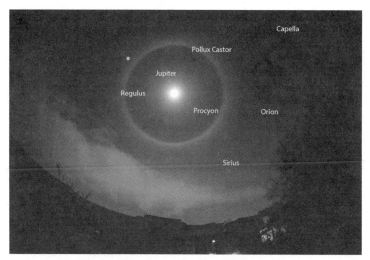

FIGURE 2.6 Moon halo on the evening of March 29, 2015. Several bright stars, Jupiter, and the constellation Orion are labeled. Theoretically, a halo around the Sun or Moon should have the spectral range of colors red, yellow, green, blue, and purple, but usually most of it appears white.

FIGURE 2.7 Sun halo.

a red color, while most of the halo is whitish. The most common high-level clouds are wispy cirrus clouds that are typically found at heights of twenty thousand feet or higher. Cirrus clouds are composed of ice crystals from the freezing of water droplets. Exactly how the crystals form is not fully understood. These crystals are about twenty microns in diameter, so small that fifty of them side by side would measure only one millimeter long.

LUNAR ECLIPSES

Lunar eclipses always occur at full Moon when Earth is between the Moon and the Sun and Earth's long shadow is cast upon the Moon (figure 2.8). During the *penumbral* phase of a lunar eclipse, the Moon enters a region of partial shadowing, the **penumbra**, and is darkened somewhat. Later when the Moon enters Earth's *umbra*, Earth's fully shadowed region, it slowly becomes much darker as it moves deeper into the umbra. If the Moon only partially enters Earth's umbra, there is partial shadowing along a portion of the Moon, and a *partial lunar eclipse* occurs. Sometimes, the Moon only enters Earth's penumbra, in which case it is only slightly darkened. Some penumbral eclipses of the Moon are quite gorgeous, while others are unremarkable and difficult to distinguish from the uneclipsed Moon. It is always worthwhile to observe any eclipse, as you might be pleasantly surprised by what it shows.

In a total eclipse the Moon gets progressively darker as it moves deeper into Earth's umbra. Totality begins when the Moon is completely within the umbra, and sometime thereafter maximum totality occurs. Theoretically, the Moon should disappear during totality because it is completely covered by Earth's shadow, but that rarely happens. Instead, the Moon turns a gorgeous copper color due to sunlight that is scattered by Earth's atmosphere and somewhat focused onto the Moon. Blue light is scattered more than red light, which produces a red or copper-red Moon. This is sometimes called a *blood Moon*, a popular term with no astronomical significance.

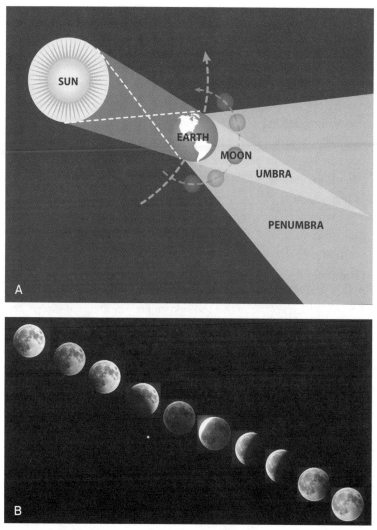

FIGURE 2.8 (a) Alignment of the Sun, Earth, and the Moon during a total eclipse of the Moon. Sizes and distances are not to scale. Drawing by Debra Bowles. (b) Total eclipse of the Moon on January 20, 2019, as viewed from Tucson, Arizona. This image is a mosaic of the initial umbral phase of the eclipse (left), totality (center), and the later umbral phase of the eclipse (right). This black-and-white image does not show the copper glow of the Moon at totality.

Totality ends when the Moon begins to leave Earth's umbra. The Moon gradually brightens as it leaves the umbra and enters the penumbra, basically brightening in reverse order to its earlier darkening in the first part of the eclipse. The complete cycle for a total eclipse of the Moon lasts for more than five hours, and totality can last for up to nearly two hours. Lunar eclipses are not uncommon, taking place roughly every three to four years. It is somewhat unusual for a given location to see a complete eclipse from the beginning penumbral phase through totality to the ending penumbral phase. The Moon often either rises after the eclipse is in progress or it sets as the eclipse is ongoing.

Totality for a lunar eclipse is quite beautiful, but it does not rival that of an eclipse of the Sun. *A total eclipse of the Sun is an absolute stunning sight not to be missed in one's lifetime if possible.* Lunar eclipses do have the advantage, however, in that totality is visible over a much larger swath of Earth's surface than a **solar eclipse** totality, which is limited to a 100–150-mile-wide strip running over Earth's surface. Solar eclipse totality is usually limited to only thirty seconds to a couple of minutes, up to a theoretical maximum of seven and a half minutes, much less than the totality of a lunar eclipse. You are even likely to see a portion of a lunar eclipse totality in inclement weather, as the Moon peeks through the clouds from time to time. In any case, it is a good rule not to miss an eclipse, whether solar or lunar, as they are wondrous events.

3

The Planets

STARS AND PLANETS

WHAT IS the difference between a star and a planet? A lot! To the naked eye, they look the same, but the traditional five naked-eye planets, Mercury, Venus, Mars, Jupiter, and Saturn, are a lot brighter than most stars. Stars are giant balls of hot glowing gas. The Sun is the nearest star. We could not exist without it. Most of the naked-eye stars visible at night are larger and brighter than the Sun, but they are so far away they look like tiny points of light even in the largest telescopes. Stars generate their enormous energy from nuclear reactions deep in their interiors. Planets are small bodies (compared to stars) that circle around parent stars. No solar system planet has internal nuclear reactions, and almost all of its energy is received from the Sun.

The planets constantly change position against the starry background as they orbit around the Sun (figure 3.1). The planets in our solar system are relatively close to us, while the nearest star is more than four **light years** away. Most stars are even much further away. From our perspective, the stars appear "**fixed**," while the planets are "wanderers" in the sky. The two closest planets to the Sun, Mercury and Venus, in that order, move the fastest, and we can easily see a change in their position from day to day.

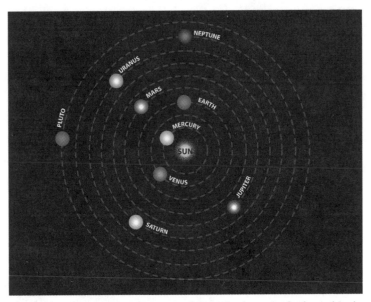

FIGURE 3.1 The solar system as viewed from far above the Sun's north pole. This is not to scale, and the orbit of Pluto goes inside that of Neptune for a portion of its orbit. Drawing by Debra Bowles.

How can you recognize a planet? It's easy once you get some experience. The best way is to have someone point out a particular planet to you, and then to try to follow it from day to day. Since all of the planets orbit around the Sun with varying speeds, you can see them change position as they slowly move relative to the far distant stars. Each planet has its own distinct color and brightness, and planets do not twinkle. Planets are close enough that we can see their disks without much optical magnification, making them less affected by the unsteadiness of the atmosphere, which causes the tiny points of light from stars to twinkle.

When I was growing up in the 1950s and 1960s, the planets consisted of Mercury, Venus, Earth, Mars, Jupiter, Saturn, Uranus, Neptune, and Pluto, in order from the Sun outward (figure 3.1). Nowadays, Pluto is a "**dwarf planet**" as defined by the IAU in 2006. I still consider Pluto a planet. Pluto is quite faint. You need a good-sized amateur telescope to see it.

Mercury, Venus, Mars, Jupiter, and Saturn are all bright enough to follow easily in the sky. They were known to the ancients and were actively observed by them. Uranus is sometimes bright enough (just barely) to be visible to the naked eye. It can be seen through binoculars and a small telescope, while Neptune is fainter and requires a small telescope for viewing it.

EVENING STAR, MORNING STAR

Evening star and **morning star** are lyrical names that evoke poetic thoughts. They sound magical, but such terms really don't have any astronomical significance. An evening star is a bright planet or star, usually Venus, which is prominent in the evening sky after sunset. When Venus is visible after sunset, it is in the western sky. Jupiter is quite bright and can act as an evening star if Venus is not present in the sky. In this case, Jupiter is usually in the western sky after sunset, but it could, for the sake of argument, be quite prominent in the eastern sky as twilight darkens in the west after sunset.

A morning star is, likewise, a bright star or planet (usually Venus) prominent in the predawn sky. In this case, Venus would be quite bright in the eastern twilight sky. If Jupiter is the morning star, it most likely would be bright in the predawn eastern sky, though for the sake of discussion it could be prominent in the western sky as the dawn twilight develops in the east.

On rare occasions, Mars will be as bright as Jupiter and will even slightly exceed it in brightness. In this case Mars can act as an evening star or morning star if Venus is not present. Venus is the brightest planet by far, considerably exceeding Jupiter and Mars at their brightest. The brightness of Venus is only exceeded by the Sun and the Moon. Jupiter is otherwise quite bright and is usually only exceeded in brightness by Venus. On those very rare occasions when Mars is brighter than Jupiter, Mars's distinct red color would lend even more of an allure to its being an evening star or morning star.

Is a star ever a morning star or evening star? Sometimes. At a very dark-sky location, if Venus and Jupiter are not in the morning or evening sky, Sirius, the brightest true star in the sky, may be considered a morning star or an evening star. Sirius is very bright, and if it does not have Venus or Jupiter to rival it, Sirius can quickly get one's attention in a dark-sky location. It is often brighter than Mercury, Mars, or Saturn, but it is never anywhere near as bright as Venus or Jupiter.

ELONGATION

Planetary elongation is the distance in degrees of a planet from the Sun. The term elongation is almost exclusively used for Venus and Mercury, which lie inside Earth's orbit (figure 3.1). Sometimes, Mercury and Venus are referred to as the **inferior planets**, while Mars, Jupiter, Saturn, Uranus, and Neptune are referred to as the **superior planets**. We are not making a value judgment here but merely noting that the orbits of Mercury and Venus are inside that of Earth, and the orbits of the other planets are outside that of Earth. Venus may be called an inferior planet in technical terms, but it is a superior delight for all who observe it.

When Mercury and Venus are at their greatest eastern elongations, they are visible in the western twilight after sunset. When they are at their greatest western elongation, they are visible in the eastern twilight sky before sunrise (figures 3.2a and 3.2b).

CONJUNCTION

A close grouping in the sky of two or more planets and/or the Moon is called a planetary **conjunction**. Usually, this means two or more bodies are within a few degrees of each other (figures 3.3a and 3.3b). Sometimes, they may lie within less than a degree. Conjunctions are fun to observe, and some planetary lineups are

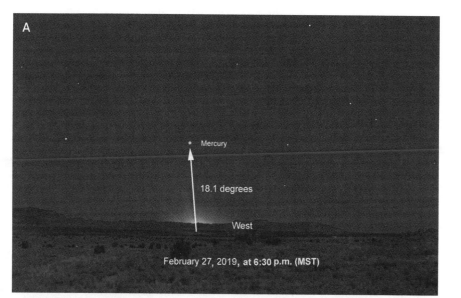

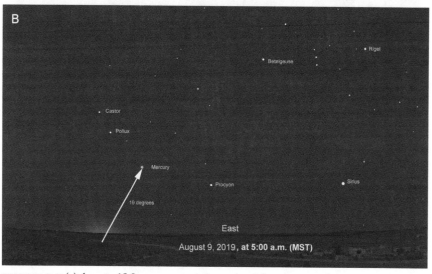

FIGURE 3.2 (a) A typical Mercury eastern elongation. Mercury is 18.1 degrees east of the Sun after sunset on the evening of February 27, 2019. (b) Mercury at 19 degrees western elongation on the morning of August 9, 2019, before sunrise. Several bright stars are labeled. This is a typical Mercury western elongation.

quite rare. However, they have no astronomical importance, as they are chance lineups by bodies that appear close to each other from our perspective but which are millions of miles away from each other.

Another, more technical definition of conjunction is when a solar system body (a planet, **comet**, or **asteroid**) has the same right ascension (the astronomical term for longitude in the sky) as the Sun as viewed from Earth. This means that if we look at the three bodies from above or below the plane of Earth's orbit (the ecliptic), they are in a straight line (figure 3.3c).

TRANSIT

If a planet (Mercury or Venus), comet, or asteroid is between Earth and the Sun, it is said to be in inferior conjunction. In rare cases, the body might cross directly in front of the Sun (transit the Sun). Since Mercury and Venus are small compared to the Sun, when they cross directly in front of the Sun, they appear as a small dark disk silhouetted by the bright background Sun. Any comet or asteroid that might transit in front of the Sun is usually so small it will not be seen. Transits of Mercury are uncommon, occurring thirteen or fourteen times per century. The most recent transits of Mercury were May 9, 2016, and November 11, 2019 (figure 3.3d). The next one will be on November 13, 2032. Transits of Venus are extremely rare. The last one occurred in 2012, and the next ones will not occur until 2117 and 2125 (figure 3.3e).

OPPOSITION

If the planet (Mars, Jupiter, Saturn, Uranus, or Neptune), comet, or asteroid lies beyond Earth's orbit, it is said to be in *superior conjunction* when the Sun, Earth, and the planet are in a straight line. The planet or object is also said to be in *opposition* when it

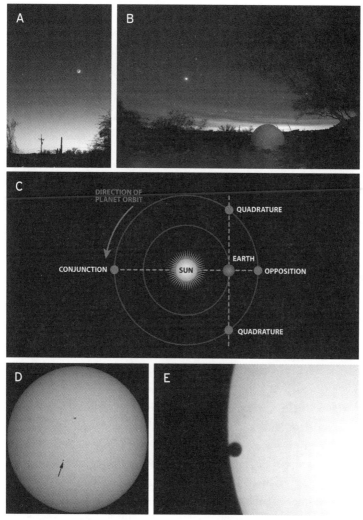

FIGURE 3.3 (a) Conjunction of the crescent Moon (bottom), Saturn (top left), and Jupiter (top right) in the western sky after sunset on December 16, 2020. (b) Conjunction of the Moon (bottom), Venus (above the Moon), and Jupiter (above Venus) on the evening of June 19, 2015. Regulus in Leo the Lion is above Jupiter. (c) Diagram illustrating the terms conjunction, **opposition**, and quadrature as if viewing the solar system from above Earth's orbital plane (the ecliptic). This would be a position far above Earth's North Pole. Drawing by Debra Bowles. (d) **Transit** of Mercury (arrow) across the face of the Sun on May 9, 2019. Several small sunspots are evident on the Sun. (e) Close-up image of Venus transiting the Sun on June 8, 2004. Venus is just at the Sun's edge near the end of the transit.

is on the opposite side of Earth from the Sun; that is, it is 180 degrees away from the Sun (figure 3.3c). This is often the best time to view the object. If the angle between the planetary body, Earth, and the Sun is a right angle (90 degrees), the body is said to be at *quadrature*.

INDIVIDUAL PLANETS AND OTHER SOLAR SYSTEM OBJECTS

MERCURY

Mercury is ever fleeting. It is the closest planet to the Sun and goes completely around the Sun every eighty-eight days (table 3.1). It is an elusive object and is not often recognized by the casual observer. Mercury never gets farther than 28 degrees from the Sun, and its daily movement through the sky is the fastest of all of the planets. It takes a little effort to see Mercury because it remains so close to the Sun. It is never far from the western horizon after sunset or from the eastern horizon before sunrise.

Nevertheless, Mercury is bright and easy to see if you have a clear horizon and know where to look. Its change in position is obvious from one day to the next. Mercury is actually a delightful naked-eye sight, but it is disappointing through a telescope. It is relatively small and, because it is always close to the horizon, the steadiness of the atmosphere ("**seeing**") for viewing Mercury is usually poor. This means a telescope's high power produces a worthless, jumpy, blurry image. Mercury does go through phases like the Moon and Venus do, and these are visible through small telescopes when seeing conditions are good.

Interestingly, Mercury is the smallest planet, with a diameter of 3,032 miles (4,879 kilometers). It is just slightly larger than the Moon, which has a diameter of 2,159 miles (3,475 kilometers). Physically, Mercury is very similar to the Moon, as it has many craters and no atmosphere. It has no moons of its own.

TABLE 3.1 THE PLANETS

	DIAMETER	ORBITAL YEAR	DAY	AVERAGE DISTANCE FROM THE SUN	MASS COMPARED TO EARTH'S
Mercury	3,032 mi (4,879 km)	88 Earth days	176 Earth days	36,000,000 mi (57,900,000 km)	0.06
Venus	7,521 mi (12,104 km)	224.7 Earth days	116.75 Earth days	67,200,000 mi (108,200,000 km)	0.82
Earth	7,926 mi (12,756 km)	365.24 days	23 hours, 56 minutes with respect to starry background; 24 hours with respect to the Sun	92,956,050 mi (149,600,000 km)	1
Moon	2,159 mi (3,475 km)	—	—	—	0.01
Mars	4,221 mi (6,792 km)	687 Earth days (1.88 years)	24.6 hours	141,600,000 mi (227,900,000 km)	0.11
Jupiter	88,846 mi (142,984 km)	4,331 Earth days (11.86 years)	9.9 hours	483,700,000 mi (778,600,000 km)	318
Saturn	74,897 mi (120,536 km)	10,747 Earth days (29.4 years)	10.7 hours	889,800,000 mi (1,433,500,000 km)	95.2
Uranus*	31,763 mi (51,118 km)	30,589 Earth days (83.75 years)	17.2 hours	1,781,500,000 mi (2,875,000,000 km)	14.5
Neptune*	30,775 mi (49,528 km)	59,800 Earth days (163.7 years)	16.1 hours	2,805,500,000 mi (4,495,100,000 km)	17.1
Pluto*	1,476 mi (2,376 km)	90,560 Earth days (247.9 years)	153.3 hours	3,670,000,000 mi (5,906,400,000 km)	0.0022

Source: https://nssdc.gsfc.nasa.gov/planetary/factsheet/.

There are five naked-eye planets visible in the night sky: Mercury, Venus, Mars, Jupiter, and Saturn.

* Uranus and Neptune are large distant planets that are not ordinarily visible to the naked eye. Technically, Pluto is a dwarf planet.

VENUS

I call Venus ever brilliant because it is so bright and gorgeous, a brilliant white "star" shining in the western evening sky after sunset or in the eastern morning sky before sunrise. Venus easily shines through **light pollution**, the bright Moon, and even thin clouds. It always gives one much pleasure to behold. Venus is the second planet from the Sun after Mercury (figure 3.1). The third planet from the Sun is Earth. Venus sticks around in the sky a lot longer than Mercury and it gets farther from the Sun, up to 47.8 degrees away, but it does move fairly rapidly through the sky from one day to the next.

Unfortunately, Venus is a big disappointment when viewed through a telescope. It has a very thick cloud cover with no easily distinguishable features and no moons. Venus does go through phases like the Moon, and these are visible through a small telescope or binoculars (figure 3.4). However, for most amateur astronomers, Venus is best enjoyed with the naked eye.

VENUS IN THE DAYTIME

Venus is so bright that it is said to be visible even during the day if it is not too close to the Sun. The trick is that you need to know

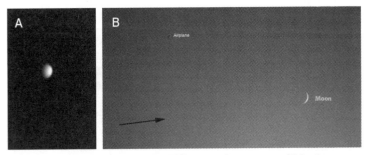

FIGURE 3.4 (a) Telescope view of Venus on the evening of March 27, 2015, when it was in a gibbous phase. The surface of Venus is covered by thick clouds that are mainly featureless. (b) Daytime Venus. On January 29, 2009, Venus (arrow) was close to the 3½-day-old Moon in the west. This picture was taken at 5:40 p.m. (MST), fifteen minutes prior to sunset at 5:55 p.m.

exactly where to look and look at that very spot. I have tried it many times without much success, though I have caught fleeting glimpses of Venus in the daytime. It is best to situate yourself such that the Sun is hidden by a building or large tree and Venus is very close to a distant object, such as a church steeple or a radio tower. Those who have practiced this sort of observing have seen Venus regularly in the daytime.

MARS

Mars has a distinctive red/orange color and appearance. Once you have become familiar with Mars, it is easy to follow it from day to day and month to month. Sometimes, when Mars is farthest from Earth, it is not very bright, but it is always distinctly red or orange. Mars varies in color depending on its seasons and on the amount of dust in its atmosphere from large dust storms. Also, its color appears to change depending on the quality of the night sky and Mars's elevation above the horizon.

Jupiter has a white color and usually is brighter than Mars. But when Mars comes closest to Earth every few years, its brightness can slightly exceed that of Jupiter. Even at its faintest, Mars is easily visible in most light-polluted skies if you know where to look. When it is really bright, Mars is quite dramatic due to its red color.

Mars does have remarkably interesting surface features, including northern and southern polar ice caps and greenish-gray markings, as well as dust storms and cloud formations (figure 3.5). Unfortunately, these are not visible through binoculars. An excellent telescope and a very steady atmosphere (good "seeing") are required to have any chance of enjoying Mars's surface features.

FIGURE 3.5 Telescope view of Mars on the night of October 22, 2005. The south polar cap is evident (south is at the bottom), as well as numerous markings on the planet. Haze and clouds are over the north polar region (top).

ASTEROIDS

Asteroids are sometimes called **minor planets**. Mostly, they are tiny rocky bodies left over from the formation of the solar system about 4.6 billion years ago. Asteroids range in size from that of Vesta, the largest at 326 miles (525 kilometers) in diameter, to less than ten feet across. All asteroids have odd, irregular shapes. Individually, they do not have enough mass to form into a sphere like larger bodies (the Sun, planets, dwarf planets, and large planetary satellites) in the solar system have. No asteroid is bright enough to be seen easily with the unaided eye. A few of them, including Vesta, are sometimes bright enough to be followed with binoculars, but most require telescopic observation, and they all look like points of light, being too small to show any disk shape through amateur telescopes.

The total asteroid count is over one million, with the vast majority being quite small and occupying the space between the orbits of Mars and Jupiter in the *asteroid belt*. Several asteroids have been visited by spacecraft, and currently asteroids are a subject of much astronomical research and study.

Ceres, with a diameter of 592 miles (952 kilometers), is the largest object in the asteroid belt. It was once considered an asteroid, but because it is nearly spherical and so much larger than any other asteroid, it has been reclassified as a dwarf planet along with Pluto, Haumea, Eris, and Makemake, the latter three of which lie far beyond Pluto in a region known as the **Kuiper belt**. When I grew up, Pluto was a planet, and Ceres was an asteroid. Haumea, Eris, and Makemake had not yet been discovered. For the purposes of this book, I still consider Pluto a planet and Ceres an asteroid. No matter what we call these objects, they are most interesting bodies that deserve further professional study.

JUPITER

Jupiter is the fifth planet out from the Sun, and it is by far the largest planet in the solar system (figure 3.1; table 3.1). Its mass is

FIGURE 3.6 Jupiter on March 25, 2015, at approximately 9:55 p.m. (MST). Note the cloud bands on Jupiter and the Great Red Spot (dark oval) below center.

two and a half times that of all of the rest of the planets combined. It is always fun to look at Jupiter and wonder how surprised Galileo was when he turned his primitive telescope to Jupiter for the first time in 1609. His discovery of four large satellites (now called the Galilean moons in his honor) circling Jupiter in part led to a fundamental change in the way mankind looked at itself and its place in the solar system. Galileo's observations of the Moon, Venus, the Sun, and the **Milky Way** with his telescopes, coupled with his writings, helped start the modern scientific revolution that continues to this day.

Jupiter has many moons, most of which are too small and faint to be seen with ordinary telescopes. However, its Galilean moons (Io, Europa, Ganymede, and Callisto) are visible through the smallest telescopes and are even visible through large binoculars, especially if the binoculars are held steady by a tripod. A small telescope also shows the multiple parallel cloud bands in Jupiter's atmosphere and sometimes Jupiter's Great Red Spot, which is a large storm in Jupiter's atmosphere (figure 3.6). It has been observed for at least two hundred years, and its size and color have changed from time to time.

SATURN

Saturn is famous for its large, bright rings and is the second-largest planet in the solar system. It is the sixth planet out from the Sun after Mercury, Venus, Earth, Mars, and Jupiter (figure 3.1; table 3.1). For many people, Saturn ranks as the number one astronomical sight due to its fabulous rings (figure 3.7). They are visible even through a small telescope at low power. Saturn also has cloud bands and many moons. The largest and brightest moon of Saturn is Titan, which has a distinct red color and is visible through a

FIGURE 3.7 Saturn on March 31, 2007. Note its gorgeous rings and cloud bands.

small telescope. Saturn has a yellow or yellow-white color, making it easily recognizable, and it is reasonably bright, though nowhere near as bright as Venus or Jupiter. Jupiter and Saturn travel around the Sun much more slowly than Earth, and they seem to stay in the same part of the sky for long periods. Even so, close observation of them from day to day shows their slow movement against the starry background.

URANUS, NEPTUNE, AND PLUTO

These planets are mentioned here for completeness. They have been minimally mentioned in Sky Spy columns, as they cannot be seen with the naked eye except for Uranus, which can get bright enough to be visible to the unaided eye at a dark-sky site. Uranus is usually visible through binoculars, though you must know precisely where to look. Its blue-green color aids in its detection. Uranus was discovered by William Herschel (1738–1822), who first observed it on March 13, 1781, and initially reported it to be a comet before determining it was a planet. It had been seen by other observers prior to Herschel, but it was mistaken for a star.

On September 23, 1846, Johann Gottfried Galle (1812–1910) discovered Neptune at the Berlin Observatory. He used the calculations of Urbain Le Verrier (1811–1877), who predicted an outer planet beyond Uranus based on observations of small variations between the predicted positions for Uranus and its observed positions. Similar calculations were performed slightly earlier by the British astronomer John Couch Adams (1819–1892), and there has been a lingering controversy as to who should get the credit for the discovery of Neptune. Nowadays, the credit is generally given to all three astronomers, though it is Galle along with his assistant Heinrich Louis d'Arrest (1822–1875) who first recognized Neptune through a telescope.

Neptune is a giant planet exceeded in mass only by Jupiter and Saturn. It is too far away to be seen without optical aid. Through a small telescope Neptune looks like a tiny blue or blue-green disk. Neptune has seventeen times the mass of Earth and a diameter of nearly four times that of Earth. It is the farthest "planet" from the Sun now that Pluto is no longer considered a full-fledged planet.

Pluto was discovered on February 18, 1930, by Clyde Tombaugh (1906–1997) at Lowell Observatory in Flagstaff while examining photographic plates he took as part of a sky survey to find a planet beyond the orbit of Neptune. Pluto was a major discovery and was listed as a planet for many years. In 2006, it was reclassified as a dwarf planet. To me, Pluto will always be a planet, as I was fortunate enough to get to know Clyde Tombaugh before he passed away. Even as a dwarf planet, Pluto is a very interesting body at the edge of the inner solar system. Pluto takes 248 years to go around the Sun and is about two-thirds the size of the Moon. It even has five moons of its own, Charon, Styx, Nix, Kerberos, and Hydra in order of distance from Pluto. Pluto is so cold that nitrogen and other gases are frozen on its surface.

Pluto is in the same class as Ceres, previously the largest asteroid. Pluto has a diameter of 1,476 miles (2,376 kilometers), making it considerably larger than Ceres, but Pluto is smaller than the Moon and Mercury. In the Kuiper belt, a region of primitive objects left over from the formation of the solar system, there are several objects that rival Pluto in size, the largest being Eris with a diameter of 1,445 miles (2,326 kilometers). While Pluto is minimally larger than Eris, it has a slightly smaller mass.

THE KUIPER BELT AND OORT CLOUD

The Kuiper belt (or Edgeworth–Kuiper belt) is a vast region in the cold, outer reaches of the solar system beyond Neptune. It was first postulated to exist in the 1940s by Kenneth Edgeworth (1880–1972) but was not appreciated until 1951 when Gerard Kuiper (1905–1973), a famed twentieth-century astronomer, predicted its existence. There are probably millions of small, mostly icy bodies in the Kuiper belt. Larger bodies are also present, such as Pluto, which is the largest known Kuiper belt object in size, though the dwarf planet Eris is slightly smaller than Pluto and has a larger mass. These objects contain varying amounts of rock, water ice, and other "icy" compounds such as frozen ammonia and methane. They are collectively known as *Kuiper belt objects (KBOs)* or *trans-Neptunian objects (TNOs)*. Over two thousand TNOs have been discovered, and there are undoubtedly millions more.

The Kuiper belt is the source of many comets, as is the **Oort cloud**, the most distant region of the solar system. The objects in the Kuiper belt are thought to lie mostly in the same plane around the Sun. The Oort cloud objects are much more distant than KBOs and are thought to lie in a vast spherical shell surrounding the whole solar system. The existence of the Oort cloud was proposed by the Dutch astronomer Jan Oort (1900–1992) in 1950, though it was also suggested by the Estonian astronomer Ernst Öpik (1893–1985) in 1932, and it is sometimes known as the Öpik–Oort cloud.

Because the Oort cloud is so distant, no specific object has been discovered in it. It is thought to be the source of some comets, and it probably contains millions if not trillions of objects due to its vast volume.

While the Kuiper belt and Oort cloud are interesting concepts, they are not relevant for everyday naked-eye, binocular, and small-telescope observing. They are mentioned here for completeness and to give one an overall sense of the solar system and its components.

4

Stars

INTRODUCTION

S TARS ARE gigantic balls of hot gas, almost all hydrogen with small amounts of helium and traces of oxygen, nitrogen, and carbon. Very tiny amounts of other elements, such as phosphorus, nickel, iron, and titanium, can be found in many stars. Gravity makes stars contract, which heats their gas to enormous temperatures, high enough to start nuclear reactions in their interiors that then produce vast amounts of energy that we see as starlight. No matter what temperature scale you use, Celsius, **Kelvin**, or Fahrenheit, the interiors of stars have temperatures of millions of degrees.

Two forces within a star constantly compete with each other. On its own, gravity would cause a star to collapse on itself, while the pressure caused by the energy a star radiates would cause it to expand. The struggle between these two forces produces a steady state for a long while, which gives the star a somewhat fixed brightness and size. This is what we observe for most of the stars we see in the night sky.

Nonetheless there are many processes by which a star can change its size or brightness over hours or days, which we then observe. When a massive star considerably larger than the Sun finally exhausts the nuclear processes in its core, it collapses upon itself and then bounces back, "exploding" as a **supernova**. Massive stars live only for a few million years until their nuclear processes run out. They use up nuclear materials much faster than more modest-sized stars like the Sun. The Sun has enough fuel to perk along with no major change for several billion years. That is good news for us. Much smaller stars are much more delicate in how they use their nuclear fuels and can live for tens of billions of years.

Except for the Sun, all stars are very far away from us. The most common and simplest way to describe stellar distances is in terms of light years, i.e., the time it takes light from the star to travel to us. The nearest star is Proxima Centauri, lying a "mere" 4.24 light years away. A light year is equivalent to approximately 5.88 trillion miles.

Proxima Centauri is a faint companion star to Alpha Centauri, the third-brightest star in the sky. In fact, Alpha Centauri is two stars that take 79.9 years to revolve around a common center of mass. Proxima Centauri is a probable third companion in this system and happens to lie slightly closer to us than the other two stars. The Alpha Centauri system is quite spectacular but unfortunately is too far south to be seen above 30 degrees north latitude.

Sirius, the brightest star, is 8.6 light years away. It is larger and brighter than the Sun, but its top billing as the brightest star is mostly because it is so "close." Except for the Sun, no star is close enough such that we can see its stellar disk, even with the largest of telescopes. Stars are just points of light because of their great distances. They may look like disks or tiny blobs through a telescope, but that is an optical effect and is not a star's actual disk such as the one we see for the Sun when viewed through a telescope with a proper solar safety filter.

THE SUN

The Sun is at the center of the solar system (figure 3.1). It provides nearly all of the energy on Earth, and without it we could not exist. The Sun is a nearly perfect sphere of very hot ionized gas with a diameter of about 864,000 miles (1,390,000 kilometers). At its very center six hundred million tons of hydrogen are fused into helium every second, releasing enormous amounts of energy. It is estimated that the core of the Sun has a temperature of 15,770,000 Kelvin. What appears to be the "surface" of the Sun is the photosphere, where visible light can escape from the Sun. The photosphere has a temperature of approximately 5,800 Kelvin. Thus, the Sun is a gas sphere nearly a million miles in diameter and glowing hot at 5,800 Kelvin.

SOLAR ECLIPSES

At new Moon, the Moon is between the Sun and Earth. The Moon casts a long thin shadow into space that misses Earth most of the time. When the Moon's shadow happens to cross Earth, a solar eclipse occurs within the path of the shadow along Earth's surface (figure 4.1a). If you are directly in the shadow's path, you will see either a *total eclipse of the Sun* or an **annular eclipse** of the Sun (figures 4.1b and 4.1c). The path of totality is narrow, not more than 150 miles wide, and totality never lasts longer than seven and a half minutes, usually quite less than that. Nevertheless, a total solar eclipse is such a stunning experience that you should do your best to see at least one in your lifetime, even if you must travel several thousand miles to the path of totality.

In the case of an annular eclipse, the Sun–Moon configuration is the same, but the Moon is too distant from Earth to completely block all of the Sun (figure 4.1c). A ring (annulus) of sunlight surrounds the Moon. These eclipses are quite spectacular but do not have the stunning beauty of a total solar eclipse.

Since the Sun is not fully covered during an annular eclipse, it is absolutely necessary that you observe the Sun using a safe solar filter or safe image projection methods. See the following website for more safety information on how to view the Sun: http://eclipse.gsfc.nasa .gov/SEhelp/safety.html. Very inexpensive solar filters can be purchased online or at science centers, museums, and astronomy shops.

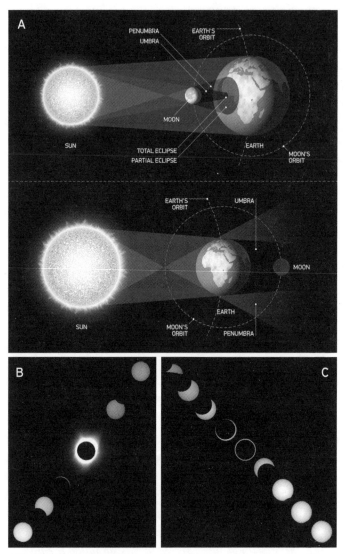

FIGURE 4.1 (a) Sun–Moon configuration for solar and lunar eclipses. Drawing by Debra Bowles. (b) Montage of a total eclipse of the Sun on August 21, 2017, as seen in Madras, Oregon. (c) Montage of an annular eclipse of the Sun on May 20, 2012, as seen in Page, Arizona.

Solar filters must be used to view the Sun during a total eclipse in all of its stages except for the few minutes of totality. Filters are required for all stages of an annular eclipse and for the entirety of a partial solar eclipse, and, of course, for any solar viewing unrelated to an eclipse.

A *partial solar eclipse* is visible in areas outside the path of totality or annularity but inside a path in which the Moon covers a portion of the Sun. Some solar eclipses are only partial, with no place on Earth lying in the path of totality or annularity.

A long-awaited "Great American Eclipse" took place on August 21, 2017 (figure 4.1b). That was the first total solar eclipse visible from the continental United States since 1979. The next one will be on April 8, 2024. The 2017 eclipse had a narrow path of totality about seventy miles wide that went across the entire country from Oregon to South Carolina. The NASA eclipse website gives a great summary of what was seen and is a good starting point for learning about eclipses in general: https://eclipse2017.nasa.gov/. The 2024 total solar eclipse will extend from Mexico into Texas and through part of the Midwest and northeastern United States before going on to Nova Scotia and Newfoundland. As the 2024 solar eclipse draws closer, be sure to check out the NASA website for more eclipse details.

STAR BRIGHTNESS (STELLAR MAGNITUDE)

Magnitude is a fancy word for star brightness. The ancient Greek astronomer Hipparchus of Nicaea (ca. 190–120 BC) created a measurement scale in which the brightest stars were defined as having first magnitude and the faintest sixth magnitude. This was popularized by Ptolemy (ca. AD 100–ca. AD 170) in his classic book the *Almagest*.

In 1856, Norman R. Pogson (1829–1891) of Oxford University proposed the modern system of magnitudes in which five magnitude steps correspond to an exact difference in brightness of a factor of one hundred. Each magnitude is 2.512 times fainter than the preceding magnitude. Very bright objects, such as the Sun (–26.8), the full Moon (–12.6), Venus (–4.9), Jupiter (–2.64), and Sirius (–1.46) have negative magnitudes.

At first glance, the magnitude is somewhat counterintuitive, but it works and has long been accepted by astronomers. Most stars vary in magnitude by a slight amount over periods of time from minutes to hours to days to years, some rather dramatically at times. Fortunately for us, the Sun is very steady in its energy output and magnitude. The planets also vary somewhat in brightness as they travel around the Sun, changing their distances from Earth.

The higher the magnitude, the fainter the star. Large professional telescopes routinely image stars of twenty-fourth magnitude and fainter. These magnitudes are "apparent" magnitudes because they represent how bright a star appears to us. They say nothing about the intrinsic brightness of the star. A very bright star far away will seem dim to us, but a truly faint star much closer will appear brighter.

First-magnitude stars are the brightest stars in the sky. The traditional list ranges from the brightest star, Sirius (magnitude –1.46), to Regulus (magnitude 1.35), the twenty-first-brightest star in the sky. The second-magnitude stars start with Adhara at magnitude 1.50.

STAR NAMES

Most of the several hundred brighter naked-eye stars have multiple names that vary in familiarity. This is well summarized on Jim Kaler's Stars website, given in the Astronomical Resources section at the end of this book. Many of these star names come down to us from myth and legend as summarized in several ancient texts, the most famous of which is the *Almagest* by Ptolemy. Others star names are of more recent origin or have been formally adopted by the IAU.

Most of the brighter stars in the sky have names with Arabic roots, the result of the fact that early Islamic astronomers working in the Middle Ages adopted Greek constellations from Ptolemy and applied their own Arabic names to stars. When Islamic astronomy texts were translated into Latin, translation and transcription errors mangled many of the Arabic names such that the ones we know today can bear little resemblance to the originals. Other star names come from ancient Greek or Latin and refer to the location or another unique characteristic of a particular star. Examples of ancient Greek, Latin, and medieval Arabic star names include Sirius ("searing" or "scorching" in ancient Greek), Polaris ("pole star" in Latin), Deneb ("tail" in Arabic) in Cygnus the Swan, and Denebola ("lion's tail" in Arabic) in Leo the Lion.

Other, more formal scientific systems were developed for star names in more recent times, including lower-case Greek letter names as proposed by Johann Bayer (1572–1625) in 1603 and specific star numbers as proposed by John Flamsteed (1646–1719) in 1712.

Bayer's system uses the Greek alphabet to name stars in order of brightness, with the Greek letter placed in front of the possessive form of the constellation's Latin name. The brightest star in a constellation is named Alpha, the second-brightest star Beta, and so forth. Vega, the brightest star in Lyra the Lyre, is Alpha Lyrae, and Sirius, the brightest star in Canis Major the Greater Dog, is Alpha Canis Majoris in the Bayer system. The names Bayer assigned to many stars remain in common use today, but the brightness order for the stars in a given constellation often does not line up well with the names assigned by Bayer. Some Beta stars are brighter than some Alpha stars in a few constellations. Since there are only twenty-four letters in the Greek alphabet, Bayer had to devise further names using upper-case and lower-case Roman letters.

The Flamsteed system builds on the naming schemes just discussed. The bright, first-magnitude stars are given their known proper names, for example Vega and Sirius, and then Greek and Latin letter names as proposed by Bayer. After that, all of the brighter stars in a constellation have their historically assigned names. The dimmer stars are then given Flamsteed numbers in numeric order according to their position in the sky going from west to east, starting with 1 for the westernmost star. Like Bayer names, Flamsteed numbers are still used today.

There are several more modern professional astronomical stellar catalogs and numbering systems. One of the more commonly used catalogs is the *Henry Draper (HD) Catalogue*, which has spectroscopic classifications and identification numbers for thousands of stars reaching far beyond naked-eye visibility. Over 250,000 stars are serially numbered in the *Smithsonian Astrophysical Observatory (SAO) Star Catalog*, while nine thousand stars visible to the naked eye are included in the *Bright Star Catalogue* and over a hundred thousand stars are listed in the *Hipparcos Catalog*. These and other formal catalogs are used by professional astronomers today.

As you can see, star naming is a complex business. The very brightest stars are mostly known by their historical names. Somewhat dimmers stars that are bright enough to be seen in most suburban skies are sometimes listed by their historical name, sometimes by their Bayer Greek letter name, or, they may be faint enough to be given a Flamsteed number, an HD number, an SAO number, or some other catalog designation. This can be very distracting and irritating for the amateur astronomer trying to look up or observe one of these stars, as their naming is not consistent from one source to another. Fortunately, here we are mainly concerned with the very brightest stars and will be using their historical names. One final note about star names: no private organization or corporation has naming rights for stars. Any star name bought through a private organization is not recognized by anyone else.

STAR COLORS

At first, stars simply appear white, but with a little experience you can see not only white but also red, orange, yellow, blue-white, and blue stars. The different colors of stars signify different properties of stars. Red and orange stars have low surface temperatures compared to the Sun's 5,800 Kelvin, while blue stars are much hotter than the Sun, with temperatures of 10,000–20,000+ Kelvin. Hot blue and blue-white stars lead fast, furious lives and use up their nuclear fuel in a few million years. Stars with temperatures like the Sun's 5,800 Kelvin amble along at a more moderate pace. The Sun has an age of nearly five billion years and several billion years of steady output still left in it.

Since stars are giant balls of hot gas with intense thermonuclear reactions deep in their interiors, they have no true "surface." The enormous energy produced by their interior nuclear reactions is conveyed to the outer parts of the star until a region is reached where visible light is no longer absorbed by the hot gas and radiates from the stellar "surface." A star's temperature and radius are measured at this "surface" so that we can compare one star with another.

Many of the bright orange or red stars visible to our naked eye are very large, old, bloated stars that have used up most of their nuclear fuel and are at the end of their lives. Some may gradually fade down to dimmer white dwarf stars, while others may undergo a violent supernova explosion that will leave a strange neutron star or a black hole in their place. In about five billion years, the Sun will run out of its primary fuel hydrogen and will swell up for a while as a bloated red giant star, then it will gradually shrink down to become a long-lived but somewhat faint white dwarf star.

FAMOUS BRIGHT STARS

There are approximately two hundred stars bright enough to be visible from many urban locations despite moderate to severe light pollution. From suburban and rural locations, many more stars are visible: probably around two thousand individual stars on a given night from a dark-sky location. This does not count the innumerable faint stars that blend in the Milky Way or faint stars grouped together in a few of the star clusters visible to the naked eye.

Jim Kaler, professor of astronomy at the University of Illinois, is one of the world's experts on stars, and his Stars website (given in the Astronomical Resources section) is an invaluable resource not to be overlooked. I refer to it often. He has an interesting list of the 172 brightest stars, many of which have been discussed in the Sky Spy columns and many of which are discussed here. The selection of which stars to individually mention here is somewhat arbitrary. Those mentioned usually are of interest because of their brightness, color, or the constellation in which they reside, or because they have interesting scientific attributes and are easy to find.

ACHERNAR

Achernar is the ninth-brightest star at 0.46 magnitude. It is very low in the southern sky from our point of view in the continental United States. I have seen it several times from southern Arizona. It is a fabulous sight hovering above the horizon (figure 4.2). It is even more amazing when you consider that Achernar is 144 light years away and at least 2,900 times more luminous than the Sun. It is six to eight times more massive than the Sun and is an elongated, egg-shaped object with a very fast rotation period of only 2.2 days. It is a strange star, indeed.

FIGURE 4.2 Achernar rising above the southern horizon as viewed from Sonoita, Arizona, on November 4, 2005, at 11:20 p.m. (MST).

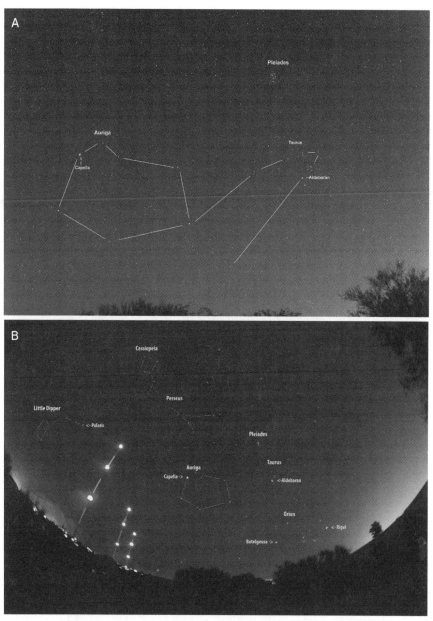

FIGURE 4.3 (a) Wide-angle view showing the constellations Taurus the Bull (which includes the Pleiades) and Auriga the Charioteer. Aldebaran is the brightest star in Taurus, and Capella is the brightest star in Auriga. (b) Extra wide angle view showing the early winter evening sky above the northern and eastern horizons. Three 300-foot radio towers provide a perspective for this nearly 180-degree view.

ALDEBARAN

Aldebaran is the brightest star in Taurus the Bull and the fourteenth-brightest star, just slightly brighter than Antares. Aldebaran is close to the Sun's apparent path in the sky, and it is regularly occulted (covered) by the Moon. Aldebaran is a reddish/orange star with a surface temperature of 4,010 Kelvin, compared to the Sun's temperature of 5,800 Kelvin. Aldebaran has a **luminosity** of 425 times that of the Sun and a diameter of about forty-four times that of the Sun.

Most of Taurus consists of a *V* configuration of stars (figures 4.3a, 4.3b, 4.4, 5.6, and 6.3b). If you look at this portion of Taurus with binoculars, you will notice many stars scattered among the brighter stars comprising the *V*. All of these stars except for Aldebaran are part of the Hyades star cluster. Aldebaran is "only" sixty-seven light years away. It sits in front of the Hyades, which lies 150 light years away. The Hyades are supposed to be the half-sisters to the Pleiades, the Seven Sisters.

FIGURE 4.4 Looking east at 7:00 p.m. on Christmas night. The bright constellations Auriga the Charioteer, Taurus the Bull (with the Pleiades), Gemini the Twins, and Orion the Hunter are rising in the early evening winter sky.

ALGOL

In the western leg of Perseus the Hero is the star Algol ("demon" in Arabic; figures 4.3b and 5.6). Algol is a close **double star**. It dims in brightness by 70 percent over roughly ten hours every 2.867 days, when a larger dimmer star passes directly in front of a smaller brighter star. Its demon-like nature has been known since ancient times, though its exact timing and mechanism are more modern discoveries.

ALHENA

At the foot of Gemini the Twins sits Alhena, the forty-third-brightest star (figure 4.4). It is quite a prominent star and would gather more fame if it were not outshone by Pollux and Castor, the "twins" in Gemini. Alhena is a double star 105 light years away. Its main star has 2.8 times the mass of the Sun.

ALPHARD

A must-see in Hydra the Water Snake is the bright star Alphard, which is the forty-seventh-brightest star in the sky (figures 4.5a and 4.5b). It can be viewed through moderate light pollution and has a distinct pale orange color. Alphard's luminosity is 946 times that of the Sun, and it has a radius fifty-eight times larger than that of the Sun. Alphard would be more spectacular if it did not lie 180 light years away.

ALTAIR

See the Summer Triangle.

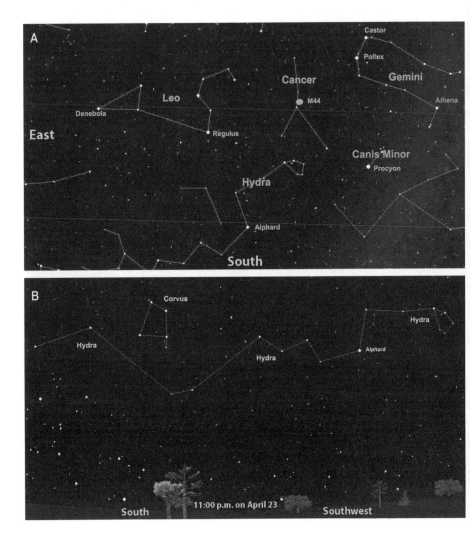

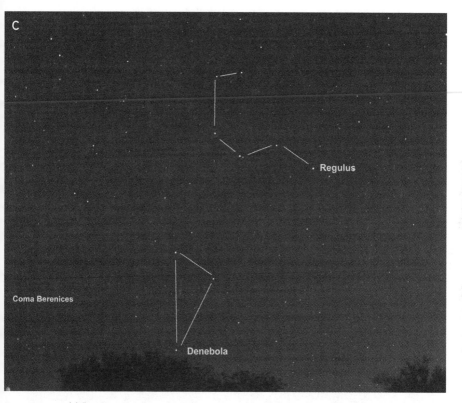

FIGURE 4.5 (a) Looking south in the early evening in midwinter. Leo the Lion, Cancer the Crab with M44 (the Beehive cluster), Gemini the Twins, Canis Minor the Lesser Dog, and Hydra the Water Snake are evident. (b) Looking toward the southwest at around 11:00 p.m. in late April. The entire extent of Hydra is evident, as well as Corvus the Crow. (c) Leo the Lion with the bright stars Regulus and Denebola. Nearby is the bright naked-eye visible star cluster Coma Berenices.

ANTARES

Antares is a bright red star, the fifteenth-brightest star in the sky. It is sometimes said to be the rival of Mars since its red color is like that of Mars, and it can be as bright as Mars when Mars is in a dim phase, distant from Earth (figures 4.6 and 5.9). Antares is at the heart of Scorpius the Scorpion.

Antares is a giant star of enormous size. It is 550 light years away and is far larger and brighter than the Sun. Antares has 15–18 times the mass of the Sun and a luminosity of 60,000–90,000 times that of the Sun. It is a cool star with a surface temperature of 3,600 Kelvin. Antares is in the last stages of its life, which will end "soon" (astronomically speaking) in a supernova explosion. That could happen tomorrow or a million years from now. It would be too far away to hurt us, but we would sure notice it.

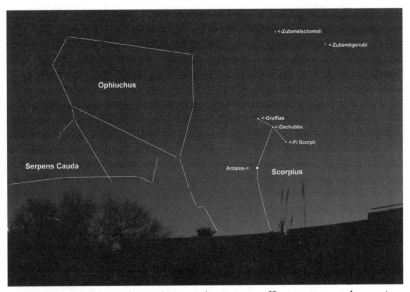

FIGURE 4.6 Looking to the southeast on the morning of January 15, 2005. Antares is at the heart of Scorpius the Scorpion. Graffias, Dschubba, and Pi Scorpii are at the head of the Scorpion. Zubeneschamali and Zubenelgenubi are extended claws of the Scorpion, though they formally lie in Libra the Scales.

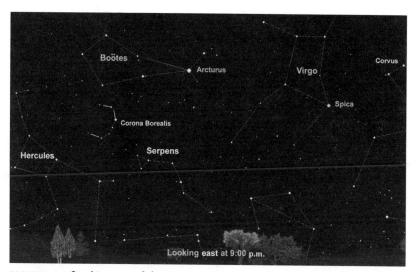

FIGURE 4.7 Looking toward the east at 9:00 p.m. in mid-spring. At this time, it is easy to contrast the red color of Arcturus with the blue color of Spica.

ARCTURUS

Arcturus is the fourth-brightest star in the sky and lies thirty-seven light years away (figure 4.7). The handle of the Big Dipper (Ursa Major) points to Arcturus. Its light was used to open the 1933 World's Fair in Chicago, because that light had left Arcturus at around the same time as the 1893 Chicago World's Columbian Exposition.

Arcturus has one and a half times the mass of the Sun and is a giant star with a diameter twenty-five times that of the Sun. It has a distinctive orange or reddish-orange color and is an old star near the end of its life. Arcturus is 113 times brighter than the Sun in visible light, but it radiates much of its energy in the infrared portion of the spectrum and emits 215 times more radiation than the Sun. It has a relatively low temperature of 4,290 Kelvin. Because its motion through the Milky Way **galaxy** is somewhat unusual, it is thought that Arcturus may have come from a small galaxy that merged with the Milky Way billions of years ago.

BRIGHT STARS OF ORION

Remembering stars names is challenging. The brightest stars in Orion are easy to see, they have interesting names and color contrasts, and they are enormous supergiant stars far larger and brighter than the Sun. The stars in Orion happen to be the brightest and largest stars in a part of the sky with giant gas clouds (**nebulae**) where dust and gas condense into newborn stars. The closest stars in Orion are hundreds of light years away, but they are bright because they are supergiant stars. Large stars are like mythical heroes who die young. Their furious lives last only a few million years.

Betelgeuse is a red star, while Rigel is a blue star (figures 4.3b, 4.4, and 4.8). The stars in Orion's "belt," Alnitak, Alnilam, and Mintaka, are blue supergiant stars. Saiph is so hot, much of its output is ultraviolet light that is not visible to human eyes. If we could see in the ultraviolet like bees do, Saiph would be even brighter. Bellatrix is unique because it is not associated with the other stars in Orion. It is much closer to us and is a foreground star.

Betelgeuse is one of the most popular stars, because of its strange name, its brightness, and its enormous size. Betelgeuse is one of the largest stars known, with a diameter of at least six hundred times that of the Sun. It has a luminosity of no less than eighty-five thousand times that of the Sun. Its distance is not known with certainty, but it is in the range of 570 light years. Betelgeuse is so large that, if it were to replace the Sun in our solar system, it would extend beyond the orbit of Mars.

Betelgeuse is perhaps ten million years old and will "soon" (astronomically speaking) end its life in a supernova explosion. When that will happen, no one knows. It could be tomorrow or hundreds of years from now. If Betelgeuse goes supernova, it will not harm us, but it will be a spectacular sight brighter than the first quarter Moon.

Rigel, the seventh-brightest star in the sky, has a blue-white color that contrasts nicely with the red color of Betelgeuse. Rigel is also a supergiant star with a diameter of approximately seventy-five times that of the Sun and a luminosity of at least eighty-five thousand times that of the Sun. It lies at a distance of 860 light years and someday will probably swell up in size to become a red supergiant star like Betelgeuse. Both stars have roughly the same mass, somewhat less than twenty times that of the Sun. One is "ballooned up" and red, and the other is more "compact" and bluer.

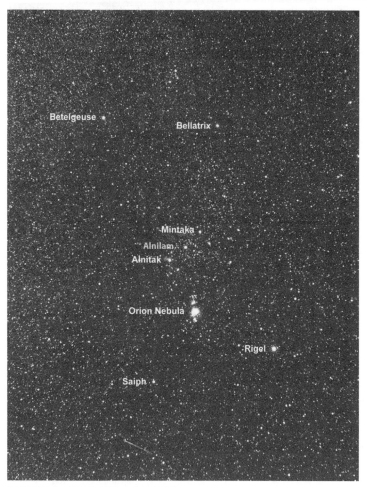

FIGURE 4.8 The bright stars of Orion.

CAPELLA

Capella sits on the northern side of the misshapen pentagon that represents Auriga the Charioteer (figures 4.3a, 4.3b, 4.4, and 5.6). Capella is yellow-white, and its name means "the she-goat," a name from Roman times. Capella is forty-three light years away and is a double star. It consists of a pair of bright stars that are closer together than Earth and the Sun and revolve around each other every 104 days. They are so close together that it takes special instrumentation or the Hubble Space Telescope to see them as individual stars. The individual components of Capella are ninety-three and sixty-four times more luminous than the Sun, and their radii are 13.6 and 8.3 times that of the Sun. Their masses are three and two and a half times that of the Sun. The combined light of these two stars plus that of possible other small companions makes Capella the sixth-brightest star in the sky. Capella is a favorite star of mine because it is so bright, colorful, and far enough north that it is visible almost all night long no matter the time of year.

CASTOR AND POLLUX, THE TWINS

Gemini the Twins is a constellation with two especially bright stars, Castor (to the north) and Pollux (to the south), the "twins" (figures 4.4, 4.5a, and 6.3b). Castor is white, while Pollux is orange-red. Nearby is the very bright star Procyon in Canis Minor, the Smaller Dog (figures 4.5a and 6.3b).

Castor is a famous "double star" used by amateur astronomers to test the optics of their telescopes. Through a telescope with good optics and sufficient magnification, Castor appears as two close, nearly equally bright stars. It is a gorgeous sight. These stars orbit around each other.

But there is more. Each of these stars is in fact a close double star that cannot be resolved by ordinary telescopic means, and there is yet more. There is a third, fainter star that is part of the Castor system. This star is also a very close double star. Thus, one of the Twins, Castor, consists of three sets of twins traveling through space together, orbiting around each other in a complex fashion.

Some double stars are traveling through space together and revolve around each other. They orbit around a common center of gravity and technically are called "binary stars." Other double stars are merely two stars that happen to lie close to each other in the sky from our point of view but have no actual connection with each other.

Pollux is just as interesting as Castor, and it is slightly brighter. Recently, it has been discovered to have a giant planet orbiting it. It is the brightest star currently known to have a planet.

DENEB AND ALBIREO

See also the Summer Triangle.

The constellation Cygnus has within it another star pattern known as the Northern Cross. The brightest star in Cygnus is Deneb, at the top of the cross (figure 4.9). Cygnus is more properly known as the Swan. Deneb is the tail of the Swan, because the "cross" and the "swan" are pointed in different directions. Deneb means "tail" in Arabic. Even though Deneb is literally at the rump of the Swan, it is a majestic star, indeed. Deneb is the nineteenth-brightest star in the sky and is one of the most luminous stars in the entire Galaxy. Deneb is so far away that its exact distance is not known, but it is thought to have a luminosity of nearly sixty thousand times that of the Sun. Deneb has a diameter of over one hundred times that of the Sun.

At the head of the Swan at the southern point of Cygnus is the wonderful star Albireo. It is not bright, but it can be easily seen in light-polluted skies. Albireo gets its notoriety from being a beautiful double star when viewed through a small telescope. The two stars that make up Albireo have contrasting golden and blue colors in a telescope, though to the naked eye they look like a single, somewhat golden star.

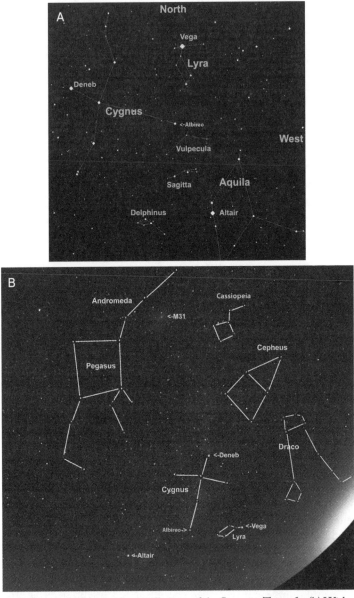

FIGURE 4.9 (a) Stars and constellations of the **Summer Triangle**. (b) Wide-angle view of the northwest early evening winter sky showing Andromeda, the Andromeda Galaxy (M31), Pegasus, Cassiopeia, Cepheus, Cygnus, Lyra, Draco, and the setting Summer Triangle of Altair, Deneb, and Vega.

DENEB KAITOS

At the western end of Cetus the Whale (or Sea Monster) is Deneb Kaitos, a fairly bright star visible even in light-polluted skies. It stands out in a lonely part of the sky south of the great square of Pegasus and northeast of Fomalhaut. Deneb Kaitos is ninety-six light years away and has a mass of three times that of the Sun, with a luminosity of 145 times that of the Sun.

DENEBOLA

Leo the Lion somewhat represents a lion at rest with its front paws outstretched toward the west (figures 4.5a and 4.5c). The front part or western part of Leo is a backward "question mark" or "sickle" of six bright stars with Regulus, the brightest star in Leo, at the bottom of the sickle. The rump or back part of Leo is a triangle of three bright stars, the brightest of which is Denebola at the easternmost point of the triangle. Most star names are difficult to pronounce and remember, but an exception to this rule is Denebola ("lion's tail" in Arabic). Denebola is the sixty-second-brightest star in the sky and, besides its brightness, it has the added advantage of a catchy, easy to remember name. It is 35.9 light years away, with a radius of 1.65 times that of the Sun, and it emits 13.8 times more energy than the Sun.

DSCHUBBA (DELTA SCORPII)

Three bright stars Graffias, Dschubba, and Pi Scorpii, from east to west, respectively, make up the head of Scorpius the Scorpion (figures 4.6 and 5.9). Graffias, Dschubba, and Pi Scorpii are all very remarkable stars, the most interesting of which is perhaps the center star Dschubba (also known as Delta Scorpii). While it has always been known as a bright star, Dschubba generally has been the fifth-brightest star in Scorpius, which is a bright constellation. Since 2000, Dschubba has brightened in an irregular fashion, and at times is the second-brightest star in Scorpius, only exceeded by mighty Antares at the heart of the Scorpion. Dschubba is surrounded by a disk of hot gas that flares at times. Dschubba is at least fourteen thousand times brighter than the Sun and five times larger. It probably has at least three companion stars.

(LONELY) FOMALHAUT

I call Fomalhaut "lonely" because it is the only bright star seemingly in the middle of nowhere, with no other bright stars anywhere close to it, particularly in an urban, light-polluted sky (figure 4.10). When I look at Fomalhaut above the sky glow over Tucson, it is by itself. Fomalhaut is a favorite star of amateur astronomers because it has a catchy name ("fish's mouth" in Arabic) and it stands as a lonely sentinel in the southern sky. Fomalhaut is twenty-five light years from us and has a luminosity of sixteen times that of the Sun. It is the eighteenth-brightest star in the sky.

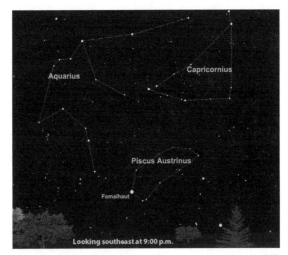

FIGURE 4.10 Looking toward the southeast at 9:00 p.m. in mid-September. Note the faint constellations Aquarius the Water Carrier, Capricornus the Sea Goat, and Piscis Austrinus the Southern Fish. Fomalhaut is the eighteenth-brightest star, but being surrounded by faint stars, it often appears by itself in light-polluted skies.

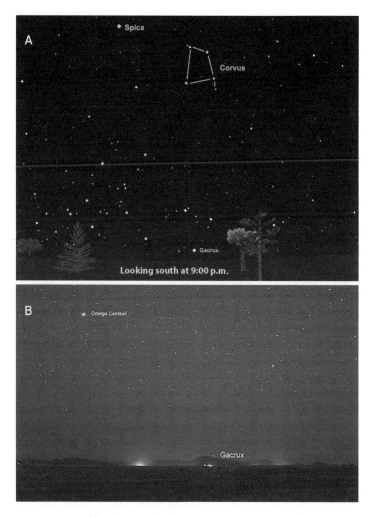

FIGURE 4.11 (a) Looking south at 9:00 p.m. in early October. If you are far enough south (below latitude 32 degrees north), you can just catch Gacrux, the northernmost star in the Southern Cross, as it skims the southern horizon. Corvus the Crow is easier to find, as is the bright star Spica in Virgo the Virgin. (b) Gacrux skimming the southern horizon as seen from near Sonoita, Arizona, on May 18, 2007, at 9:40 p.m. (MST). The glow is from car headlights, and Omega Centauri, an exceptionally large, naked-eye globular cluster, is marked. Both Gacrux and Omega Centauri are much better viewed from the Southern Hemisphere.

GACRUX

Crux the Southern Cross is a beautiful constellation with bright stars. It is so beautiful that it appears on several national flags, including those of Australia, New Zealand, and Brazil. Unfortunately, Crux is too far south to be seen in the continental United States except from the Florida Keys and the southern tip of Texas, where it is just visible above the southern horizon.

There is an anecdotal story that the topmost star in Crux is visible from Kitt Peak National Observatory, southwest of Tucson, Arizona, as the star just skims above the southern horizon (figure 4.11a). This is indeed true. The most northern star in the Southern Cross, Gacrux (Gamma Crucis), is visible from southern Arizona, south of Tucson. It is the twenty-fourth-brightest star in the sky and is very red (figure 4.11b). I have seen it from Sonoita, Arizona, forty miles southwest of Tucson, as Gacrux rose to about a degree above the horizon. I first saw it through binoculars, but it is visible to the unaided eye if you have a dark night and a very flat horizon free of clouds and haze.

KOCHAB

At the western edge of the Little Dipper's bowl is the bright star Kochab, which is just as bright as Polaris but does not receive the same share of fame. Polaris's fame lies from its chance position near the north celestial pole. It is not the brightest star in the sky, though it and Kochab are easily visible through moderate light pollution. Kochab has a slightly yellow or orange tinge. It is 126 light years away and five hundred times more luminous than the Sun.

MIRA (OMICRON CETI)

Cetus the Whale (or Sea Monster) is a faint constellation coming down to us from myth and legend. It has faint, telescopic objects of interest to amateur and professional astronomers, and it has a most interesting star, Omicron Ceti, also known as Mira, "the amazing one." Mira is the only proper named star that is often too faint to be seen. Most of the time Mira is not visible to the naked eye, but over a period of roughly 330 days it brightens enough to be easily visible in a dark sky. Sometimes it is a visible part of the constellation Cetus and sometimes not.

Mira is a very large red giant star in the last stages of its life, lying 420 light years away. It is irregularly swelling and contracting while blowing off gas and dust, and will end its life as a small, hot white dwarf star.

MIZAR AND ALCOR

Mizar, the second star from the end of the handle of the Big Dipper, forms a double star with a companion, Alcor, which is visible to the naked eye. Mizar and Alcor have been a traditional test of sight. The Arabs referred to them as "horse and rider." If you can't see Mizar and Alcor as separate stars with your unaided eye, give them a try with binoculars.

Mizar also has a bright telescopic partner that was discovered in 1650. Even more amazing is that each of the "Mizars," Mizar A and Mizar B, are extremely close double stars. There is still more. Alcor also has a faint companion star. What is uncertain is whether the Mizar and Alcor systems are traveling in space together or are simply close together in the sky from our point of view.

POLARIS

Polaris, the North Star, is so named because it lies close to the north celestial pole. Earth's complete rotation about its axis every day makes the stars appear to circle around a point in the sky, the north celestial pole in the Northern Hemisphere and the south celestial pole in the Southern Hemisphere. These points are just projections of Earth's axis into the sky. If you could stand at the North Pole, Polaris would be overhead. Because Tucson, Arizona, where I live, is at latitude 32 degrees north, Polaris lies about 32 degrees above the northern horizon. Polaris's elevation above the northern horizon is roughly equal to an observer's latitude.

At the equator, Polaris lies just on the horizon. Polaris is not at the exact polar point but lies within 1.5 degrees of it. This means Polaris actually circles around the north celestial pole once a day, but this motion is too small to be noticed by eye. Polaris has served as a navigator's benchmark for hundreds of years. If you know the sky, and if it is clear at night, you can always find north by locating Polaris. The Southern Hemisphere is not blessed with such a bright star at its pole, and direction finding at night is much more difficult.

Polaris is not the brightest star. It is the fiftieth-brightest star, but bright enough to be prominent even in light-polluted skies. Polaris is in the constellation Ursa Minor the Lesser Bear, more commonly thought of as the Little Dipper. It is at the tip of the handle of the Little Dipper, and you can use the two stars that make up the end of the bowl of the Big Dipper to guide you to Polaris (figures 4.3b and 5.3). Polaris deserves some of its fame, as it is 2,500 times more luminous than the Sun, with four times the Sun's mass, and it is bright even though it is 430 light years away.

PEACOCK STAR

In the southern part of the United States, if you have a pair of binoculars and a truly clear southern horizon free of city lights, you will be able to see the "Peacock Star" Alpha Pavonis just skimming above the southern horizon. It is the brightest star in the faint southern constellation Pavo the Peacock. Alpha Pavonis is the forty-fourth-brightest star and is a conspicuous blue-white star if viewed in the Southern Hemisphere. Its name was given to it by Her Majesty's Nautical Almanac Office (now part of the United Kingdom Hydrographic Office) in the 1930s. A navigational almanac for the Royal Air Force was created then, and it included fifty-seven stars. Since Alpha Pavonis did not have a classical name, it was assigned the name the Peacock Star. It deserves its name, as it has six times the Sun's mass and four to five times the Sun's radius and lies 180 light years away. The Peacock star is actually a close pair of stars that orbit around each other with a period of 11.8 days.

PROCYON AND GOMEISA

Canis Minor the Lesser Dog is not much of a constellation; it basically consists of two stars, the brightest of which is Procyon, the eighth-brightest star (figure 4.5a). Gomeisa is the other star in Canis Minor. Procyon appears so bright to us because it is only 11.4 light years away. Procyon has 1.4 times the mass of the Sun. It also has a tiny companion star that circles Procyon every 40.8 years. This companion, officially known as Procyon B, has a mass of 0.6 times that of the Sun. Procyon itself is officially known as Procyon A. When we view Procyon, we are seeing the combined light of Procyon A and B. Gomeisa, though appearing dimmer to us than Procyon, is a far more glorious star. It lies 170 light years away, which reduces its splendor. However, Gomeisa radiates 250 times the energy of the Sun and has a mass of three times that of the Sun.

REGULUS

Regulus is at the heart of Leo the Lion and lies about seventy-nine light years away (figures 4.5a and 4.5c). It is 150 times brighter than the Sun in visible light. Regulus is a quadruple (four star) system. The main star is spinning so fast on its axis that its poles are somewhat flattened, with the star looking like an egg lying on its side. The system of stars composing Regulus provides plenty of research material for stellar astrophysicists.

SIRIUS AND CANOPUS

Sirius ("searing" or "scorching" in Greek) is the brightest star in the sky (figures 4.12a, 4.12b, and 6.3b). Sirius is in the constellation of Canis Major the Greater Dog and is often referred to as the "Dog

Star." Sirius is approximately twenty-six times brighter than the Sun and is "only" 8.6 light years from the Sun. While Sirius is a larger, brighter star than the Sun, it mainly gets its top-dog status because it is so close to us. It also has a small companion star, Sirius B, which is a white dwarf star, a small, compact extremely hot star that is a ball of carbon and oxygen, the dying remnant of a formerly larger star now slowly cooling off.

Canopus, the second-brightest star, is close to Sirius, lying just to the south and west of it (figures 4.12b and 6.3b). Canopus is not quite as bright as Sirius and is very low in the southern sky. In fact, Canopus is not at all visible from Canada and most of the continental United States. Canopus is truly a far more magnificent star than Sirius. It is a supergiant star fifteen thousand times brighter than the Sun and has a diameter of sixty-five times that of the Sun. Even though it lies 313 light years away, Canopus's intense nuclear fires produce enough light for it to be a dominant star for those enjoying the sky in the Southern Hemisphere.

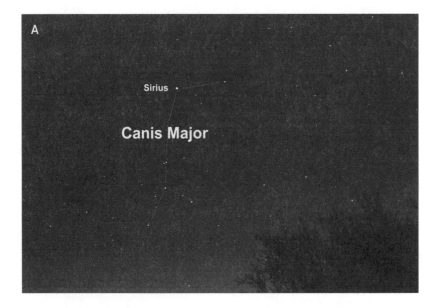

FIGURE 4.12 (a) Telephoto image of Canis Major showing Sirius. (b) Looking south over Tucson, Arizona. Sirius in Canis Major the Greater Dog is the brightest star in the sky. Nearby is Canopus, the second-brightest star.

SPICA

Spica is the brightest star in Virgo the Virgin and the sixteenth-brightest star (figure 4.7). When Spica is high above the horizon, it appears bright blue to me. It is 260 light years away and is visually 2,100 times more luminous than the Sun, though much hotter, and much of its radiation is ultraviolet light rather than visible light. Spica consists of a pair of two very close stars that orbit each other with a period of slightly more than four days. The stars have high temperatures of 22,400 Kelvin and 18,500 Kelvin. Since prehistoric times, Spica has been considered a harvest symbol because the Sun passes Spica in the sky in the fall.

Spica lies about 10 degrees south of the celestial equator (the imaginary projection of Earth's equator on to the sky). Spica also lies near the ecliptic, the part of the sky through which the Sun appears to travel over the course of the year. The Moon and planets travel within a few degrees of the plane of the ecliptic and often come close to stars that lie near it, such as Spica and Regulus in Leo the Lion. In fact, the Moon regularly occults (covers) Spica.

THE SUMMER TRIANGLE (NAVIGATOR'S TRIANGLE)

The Summer Triangle is an **asterism** of three bright stars, Altair, Deneb, and Vega. An asterism is merely a grouping of stars that is not an official constellation. Altair is in the constellation Aquila the Eagle, Deneb is in Cygnus the Swan, and Vega is in Lyra the Lyre (figure 4.9a). The Summer Triangle is reasonably easy to recognize because its stars are among the brightest in the sky.

Not only is the Summer Triangle beautiful, but each of its stars has an interesting story to tell. Altair lies seventeen light years away and is a very strange star that is rapidly rotating with an ellipsoid shape like an egg. Deneb is so far away its distance is not known with any certainty, but it is estimated to be roughly 1,425 light years from us. Even so, Deneb is the nineteenth-brightest star. Deneb is probably one of the most luminous stars in the entire Milky Way.

Vega is the fifth-brightest star in the sky and lies about twenty-five light years from us. A large disk of gas and dust surrounds Vega, with planets possibly in this disk. Jodie Foster, playing Eleanor Arroway in the movie *Contact*, made a visit to the star Vega. We cannot yet make a visit to Vega, but we can marvel at its beauty.

The Summer Triangle was popularized in amateur astronomy after the Second World War. For a long time prior to that, it was known as the **Navigator's Triangle** by U.S. military navigators who used sextants for navigation until early electronic navigational systems and later global positioning systems (GPS) took over modern navigation. This bright triangle of stars was easy for military navigators to recognize and take bearings with even in poor viewing conditions.

ZUBENELGENUBI AND ZUBENESCHAMALI

Zubenelgenubi and Zubeneschamali are two of my most favorite stars because of their striking names and because, once found, they are easily recognizable old friends north and west of Scorpius the Scorpion (figure 4.6). I must always be careful to look up their proper spelling whenever I write about them. In ancient times they were the outstretched claws of Scorpius. Zubenelgenubi is derived from an Arabic phrase meaning the Scorpion's southern claw, while Zubeneschamali is the northern claw. These two stars are not related to each other, and they are in Libra the Scales, a dim constellation otherwise not easily found. They make looking for Libra worthwhile.

5

Constellations

INTRODUCTION

CONSTELLATIONS ARE just groupings of stars used to identify a region of the sky. Some of our constellations predate written records, and most originally came from the myths and legends of Mesopotamia, Babylon, Egypt, and Greece, as codified by Ptolemy (ca. 100–170 AD), a Greek or Egyptian mathematician and astronomer who lived in Alexandria, Egypt. His great work the *Almagest* is the only surviving comprehensive text on astronomy from ancient times.

As mentioned before, the official set of eighty-eight constellations used by professional and amateur astronomers today was defined by the IAU. The boundaries of each constellation contain the stick figures we traditionally think of as constellations, as well as other stars. When the Moon or a planet crosses a constellation boundary, it is said to go from one constellation to another.

In most cases the constellations do not resemble the object, person, or animal they are supposed to represent. The individual stars within a constellation usually have no physical relationship with each other. They lie at various distances from Earth and merely happen to occupy the same part of the sky from our point of view.

In general, I prefer to use a planisphere to identify constellations and bright stars, because it is so simple to use and gives one an overview of the entire sky, something not available with apps. The constellation figures in this book are a supplemental aid to help you learn some of the more interesting constellations. They are not all inclusive, and you should use them as a complement to a planisphere, other sky charts, or mobile applications. The constellation stick figures I use are how I see the constellations and may vary a bit from those of other authors. What is important is to have fun and learn the constellations at your leisure.

While we commonly associate a constellation with a specific season, most constellations are visible for a good part of the year. To be technically correct when you label a constellation as being in a particular season, you also have to specify at what time of night it is visible. Typically, this is when it is present in the evening sky, since it is far easier for most of us to observe in the early evening than after midnight or in the predawn. However, astronomers know the sky is always interesting and enjoyable, but it is a tough taskmaster, with some of the finest heavenly events taking place after midnight or before dawn.

SELECTED CONSTELLATIONS

The Sky Spy columns and this book describe those objects and phenomena visible from the Northern Hemisphere, mainly from the continental United States. Faint, obscure, or far southern constellations are not mentioned, as they would not be visible from most of the United States. The following constellation description are centered on the larger and brighter constellations in the Northern Hemisphere.

ANDROMEDA

See Pegasus and Andromeda, and Figure 4.9b.

AQUARIUS AND CAPRICORNUS

Aquarius the Water Carrier and Capricornus the Sea Goat are faint southern constellations. Because of their faintness, they get overlooked much of the time, though Capricornus (often spelled Capricornius) is easily recognized even if it is faint. It looks like a flattened bowl with a wide mouth facing north and a pointed bottom facing south (figure 4.10).

These are ancient constellations, which were quite visible thousands of years ago when many myths and legends were being compiled. The ancients had dark skies and spent plenty of time looking at them. They used the heavens to tell time at night, to predict upcoming weather, to follow the change of the seasons, and to judge the proper time to plant crops, among many other things. Since these constellations do not look anything like what they are supposed to represent, they must have been placed in the sky for symbolic purposes. This is often a source of frustration to beginning observers. One must accept that and enjoy the skies themselves and try to imagine why our ancient forefathers designed the constellations as they did.

AQUILA

A magnificent part of the Milky Way lies in the constellation Aquila the Eagle. Altair, the brightest star in Aquila, is the thirteenth-brightest star. It is flanked on each side by a bright star, making a line of three stars with Altair in the middle. This is the head and neck portion of the "eagle." The wings and body consist of dimmer stars south of these three.

Along the western edge of Aquila, there is a large dark band that runs for a long distance from northeast to southwest through the Milky Way. This is the "Great Rift" of the Milky Way (figure 5.1). This band of relative darkness hides many intriguing areas of star formation and other astronomical phenomena of interest to professional astronomers, who use infrared imaging and radio telescopes to peer through the dust and gas.

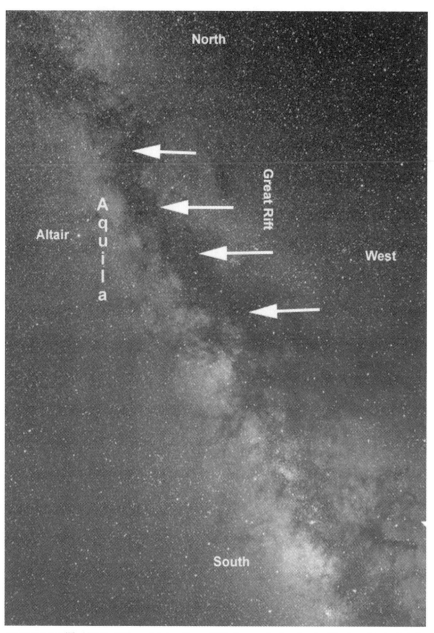

FIGURE 5.1 The Milky Way with its Great Rift (arrows). Aquila the Eagle with its bright star Altair overlies part of the Milky Way.

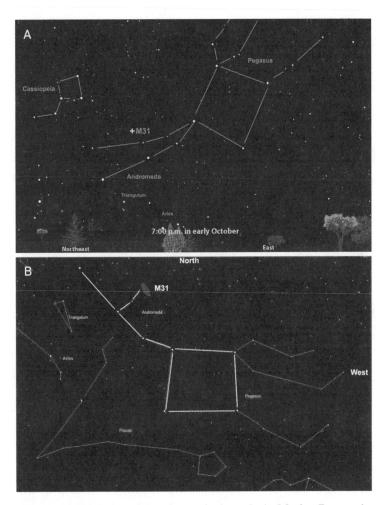

FIGURE 5.2 (a) Cassiopeia the Queen, Andromeda the Maiden, Pegasus the Winged Horse, Triangulum the Triangle, and Aries the Ram rising in the east on an early October evening. Note the Andromeda Galaxy (M31). (b) Andromeda the Maiden with M31, Pegasus the Winged Horse, Pisces the Fishes, Triangulum the Triangle, and Aries the Ram. (c) Telescope view of M31, the Andromeda Galaxy, the closest large galaxy to our own Milky Way.

ARIES AND TRIANGULUM

Aries the Ram is a member of the zodiac, but it is a small, unimpressive constellation that looks like a squished triangle (figures 5.2a and 5.2b). Aries is below Andromeda the Maiden (figure 5.2c) and is often overlooked. Even though it is small and unimpressive, Aries is relatively bright. It is easily visible once identified and can be seen through moderate light pollution.

Between Aries and Andromeda is another small triangle of stars, Triangulum the Triangle. Triangulum is fainter than Aries, but easily recognized once found. It can also be seen through moderate light pollution. Both Aries and Triangulum are ancient constellations.

AURIGA

Auriga the Charioteer is just as big and almost as bright as Orion (figures 4.3a and 4.3b). In fact, Auriga contains Capella, the sixth-brightest star. Capella is brighter than Rigel and Betelgeuse, the brightest stars in Orion, though overall Auriga is not quite as dramatic a constellation as Orion.

Auriga is a large misshapen pentagon with several extra stars along its sides or inside it. Auriga is 30 degrees directly north of Orion. Capella sits on the northern side of the pentagon of Auriga. There are many telescopic objects of interest in Auriga for amateur and professional astronomers. A portion of the winter Milky Way goes through Auriga.

THE BIG DIPPER (URSA MAJOR)
AND LITTLE DIPPER (URSA MINOR)

The Big Dipper is probably the most familiar constellation. It is part of Ursa Major the Great Bear (figure 5.3). The Little Dipper is the most recognizable part of Ursa Minor the Little Bear. These ancient constellations predate written civilization, though they did not necessarily represent bears in the lore of other civilizations.

Earth's complete rotation about its axis every day makes the stars appear to circle around the north celestial pole in the Northern Hemisphere and the south celestial pole in the Southern

FIGURE 5.3 Telephoto image of the Big Dipper and the Little Dipper. Note that the two stars in the bowl of the Big Dipper point to Polaris (the North Star) in the handle of the Little Dipper. Three radio towers north of Tucson, Arizona, are in the foreground in front of the Catalina Mountains.

Hemisphere. The Big Dipper can be thought of as a giant watch hand going completely around the north celestial pole once a day.

The Big Dipper is so well known because its stars are bright and it has an easily recognizable pattern. But there is more. The two end stars in the bowl of the Dipper point to Polaris the North Star, and the handle of the Big Dipper points to Arcturus, the very bright red star in the constellation of Boötes. Mizar, the second star from the end of the handle of the Big Dipper, forms a double star with a close companion, Alcor. The stars in the Big Dipper are moving through space with respect to our Sun. Over thousands of years the Dipper will gradually change its shape and disappear, but we have plenty of time to enjoy it now.

From Canada and most of the United States, the Little Dipper is a **circumpolar** constellation. That means we can see it all night long as it circles around the north celestial pole. Polaris is at the end of the Little Dipper's handle. There are two stars in the handle between Polaris and the bowl of the Little Dipper. The Little Dipper is dimmer and less well known than the Big Dipper, but it is bright enough to be seen through mild light pollution, and it has an easily recognizable pattern.

BOÖTES AND CORONA BOREALIS

Boötes the Herdsman looks like a pentagon of stars, to the south of which is a separate very bright star Arcturus, the brightest star in Boötes and the fourth-brightest star (figure 4.7). Arcturus is bright enough to be visible even with the Moon in the sky. It has a distinct red color and is easy to find most of the time.

The handle of the Big Dipper points to Arcturus. This relationship of Arcturus and Boötes to Ursa Major is reflected by the fact that Arcturus means "bear watcher." *Arktos* is the Greek name for bear, and the word "arctic" is derived by reference to the Great Bear circling around the northern sky. It is likely these concepts came down to us from Ice Age peoples.

There are constellations representing kings and queens as well as crowns in the sky, one of which is Corona Borealis, the Northern

Crown, which is just east of Arcturus. Corona Borealis is composed of six relatively bright stars that form a wide cup or *U*. The open portion of the *U* faces north, and the entire constellation is approximately as wide as your fist if you extend your arm toward the sky. There is also a Southern Crown, Corona Australis, which sits south of Sagittarius the Archer. It is low in our sky and not nearly as beautiful for us as Corona Borealis.

CANCER

Cancer the Crab is dim and uninspiring. Right in the middle of Cancer is the Beehive, a large star cluster (figures 4.5a and 5.4). Less commonly, the Beehive is called Praesepe (the Manger), because Greek and Roman observers described it as a manger from which two donkeys are eating. The Beehive is more commonly known as M44. The famous French astronomer Charles Messier (1730–1817) included it as the forty-fourth object in his catalog of important sky objects, but it has been known since ancient times. It is visible to the naked eye in dark skies and is a gorgeous sight through binoculars.

FIGURE 5.4 Telescope view of the Beehive star cluster (M44), which lies in the center of Cancer the Crab.

CANIS MAJOR AND CANIS MINOR

Canis Major the Greater Dog and Canis Minor the Lesser Dog are following Orion the Hunter in the sky as his faithful companions (figures 4.5a, 4.12a, and 4.12b). Canis Major is a bright, gorgeous constellation close to the winter Milky Way. It contains several bright stars and star clusters of interest to astronomers. It even contains the brightest star in the sky, Sirius. Canis Major resembles a dog standing on its hind legs with Sirius as its blazing eye. Canis Major along with Orion are traditional winter evening constellations.

To the direct east (left) of Orion and 25 degrees northeast of Canis Major is Canis Minor the Lesser Dog, the smaller faithful companion of Orion. Canis Minor is not much of a constellation, basically consisting of two stars, the brightest of which is Procyon, the eighth-brightest star. Four degrees above and to the west of Procyon is Gomeisa, the other star in Canis Minor and the 153rd-brightest star. Even though Canis Minor is small, it is easily recognizable in light-polluted skies.

CASSIOPEIA AND CEPHEUS

Cassiopeia the Queen of Ethiopia is a bright constellation that is easy to learn, while Cepheus her husband and King of Ethiopia is hard to find. Cassiopeia looks like a chair or stretched *W* or *M*, depending on its orientation in the sky (figures 4.3b and 4.9b). Cassiopeia represents a proud queen who boasted of her beauty. She and Cepheus were legendary rulers of a kingdom that is supposed to have stretched from present-day Syria down to the shores of the Red Sea. Their daughter was Andromeda the Maiden, and her husband was Perseus the Hero, all of whom made their way into the sky. Cassiopeia is an ancient constellation and comes to us in its present form from Ptolemy.

Cassiopeia gets top billing over her husband Cepheus because she is small and bright, while Cepheus is large and dim. To me, Cepheus looks like a misshapen pentagon, sort of a triangle sitting on top of a square with the triangle pointed toward the north. Cepheus needs a dark sky for proper appreciation, and it will take some work to recognize it. However, a wonderful stretch of the Milky Way runs through Cassiopeia and Cepheus with many superb star clusters and nebulae favored by amateur astronomers.

CETUS

Cetus the Whale (or Sea Monster) is a large constellation but quite faint. It is below Pisces the Fishes and Aries the Ram and partially above and to the west of Eridanus the River. These are all faint constellations. Even in dark skies, I have difficulty finding the main outline of Cetus.

Cetus is worth the effort to locate, as it is an ancient constellation coming down to us from myth and legend. It has faint objects of interest to amateur and professional astronomers, and it has a most interesting star Omicron Ceti, also known as Mira, "the amazing one." Mira is a variable star that is only visible to the naked eye some of the time. Therefore, sometimes it can be seen in the constellation and sometimes not.

COMA BERENICES

To the east and slightly north of Leo the Lion is a faint, unspectacular constellation Coma Berenices, Queen Berenice of Egypt's hair. Queen Berenice II (267 or 266 BC to 221 BC) was the wife of Ptolemy III Euergetes, the third ruler of the Ptolemaic dynasty of Egypt. In 243 BC she cut off her golden hair as a gift to the goddess Aphrodite and placed it in Aphrodite's temple to celebrate her husband's safe return from war in Syria. The hair disappeared and was later said to be carried to the heavens to be placed among the stars. The constellation Coma Berenices is composed of only three faint stars, but it contains a large cluster of about fifty fainter stars spread over 5 degrees of the sky, a gorgeous sight through binoculars or a small telescope (figures 4.5c and 5.5). This is the Coma Berenices star cluster.

FIGURE 5.5 Telescope view of the Coma Berenices star cluster.

CORVUS

Corvus the Crow is a constellation that doesn't get much press or respect, as it is somewhat small. However, Corvus is a favorite of mine. It is bright enough to be easily seen in moderate light pollution. Its somewhat misshapen square form is easily recognizable once you have found it the first time (figures 4.5b and 4.11a). If you have a dark sky, Corvus is quite bright, indeed. How Corvus is supposed to resemble a crow beats me. The ancients were certainly quite familiar with the pesky crows that ate their seeds. No doubt scarecrows are an ancient invention. Crows and their cousins the ravens are ubiquitous, and often annoying, but we must admit they are most clever and adaptable. Enjoy Corvus and let it become a familiar friend in the spring and summer night sky. If you find crows troubling during the day, at least you will enjoy this nighttime crow.

CYGNUS

Cygnus the Swan is also known as the Northern Cross, since the six brightest stars of Cygnus form a cross with the top of the cross pointing northeast and the bottom of the cross pointing southwest (figures 4.9a and 4.9b). A beautiful portion of the Milky Way runs through Cygnus. Cygnus is the constellation drawing I first saw in Miss Wilmore's book in the first grade, which started my lifelong interest in astronomy. Fortunately, Cygnus is a splendid constellation and very easy to see. Every time I look at it, I think of Miss Wilmore and those bygone days with little light pollution.

The brightest star in Cygnus is Deneb, which is at the top of the cross. However, Deneb is the tail of the Swan, because the "cross" and the "swan" point in different directions. Deneb is the Arabic word for tail, and the ancients saw Cygnus as a swan flying south along the Milky Way.

DRACO

Draco the Dragon is a circumpolar constellation for viewers in Canada and most of the United States. It is so far north in the sky that it never completely sets: it circles around the north celestial pole as Earth rotates on its axis. The head of Draco is the most southern part of the constellation. It looks like a somewhat small, squashed rectangle not too far north of the bright star Vega in Lyra the Lyre (figure 4.9b). The body of Draco is a long string of stars that runs north toward Polaris and then curves west around the Little Dipper. To me, Draco looks more like a moderately sized celestial snake rather than a fearsome dragon. Draco is an ancient constellation from myths and legends, from a time before dragons became extinct.

ERIDANUS

Eridanus the River is a meandering string of stars that starts near the bright star Rigel in the western foot of Orion and ends up 60 degrees west of Orion at Achernar, the ninth-brightest star (figure 4.2). Eridanus is dim and unspectacular, but at least it resembles what it is supposed to represent. Most of Eridanus is easy to see in a dark sky, but Achernar is a challenge because it is so far south for most of the United States. I have seen it several times from southern Arizona as it skimmed above the southern horizon.

GEMINI

Gemini the Twins is one of the bright, distinct winter evening constellations (figures 4.4 and 4.5a). Its twins are the stars Castor and Pollux, respectively the twenty-third- and seventeenth-brightest stars. Gemini sits just north of Canis Minor the Lesser Dog, northeast of Orion the Hunter, and west of Auriga the Charioteer and Taurus the Bull. Gemini serves as the radiant for the bright **meteor shower** the Geminids in December, possibly the best meteor shower of the year.

HERCULES

Hercules the Hero is a somewhat misshapen square of four stars that represents his body with stars extending to the north and south that are his arms and legs. Hercules is not one of the brighter constellations, but it is easily seen on a reasonably dark night. To the west of Hercules sits Corona Borealis the Northern Crown, and to the east of Hercules is Lyra the Lyre with its bright star Vega. South of Hercules is the underappreciated constellation Ophiuchus the Serpent Holder.

Along the western side of the "square" portion of Hercules, toward its northwest corner, is a faint density of stars that is visible to the unaided eye in a dark sky. Through a small telescope this density reveals itself to be a compact ball of thousands of stars, a globular cluster named M13, the thirteenth object described in a famous catalog of nonstellar objects compiled by the French astronomer Charles Messier in the eighteenth century. M13 is a favorite object of amateur astronomers, for both viewing and photography. It lies about 25,100 light years away and contains several hundred thousand stars.

HYDRA

Hydra the Water Snake is a large, mostly faint constellation. At least it somewhat looks like what it is supposed to represent. Hydra's head, the leading part of the constellation, is a small misshapen pentagon of five stars. It is approximately halfway between Procyon in Canis Minor the Lesser Dog and Regulus in Leo the Lion, and is just below the faint constellation of Cancer the Crab. The long body of Hydra stretches from its head toward the east, running considerably below the large constellations of Leo and Virgo the Virgin, and just below the small faint constellations of Sextans the Sextant, Crater the Cup, and Corvus the Crow (figure 4.5b). Be sure to look for Alphard, the brightest star in Hydra. Alphard can be seen in relatively light-polluted skies. It has a distinct pale orange color.

While Hydra is narrow north to south, it is the longest constellation. Enjoy Hydra. This snake does not bite.

LEO

Leo the Lion is one of the larger, brighter constellations that should become a familiar friend to you. If you consider Leo to be a lion resting on his stomach with his paws out in front, the head and chest of Leo face west and look like a backward "question mark" or sickle in the sky. This part of Leo consists of five bright stars in the sickle portion, with the very bright star Regulus acting as the handle of the sickle or the dot on the question mark (figures 4.5a and 4.5c)

Leo's rump is to the east and is a triangle of bright stars. The brightest star in Leo's rump, on its far eastern end, is Denebola ("lion's tail" in Arabic), the sixty-first-brightest star. Regulus and Denebola help make Leo the King of the Beasts.

Leo is also popular with astronomers because it contains many important telescopic objects. Some of these include relatively large, bright galaxies visible through small telescopes. Leo also gets its notoriety by being in the zodiac, that narrow band of the sky through which the Sun, Moon, and planets travel.

LEPUS AND COLUMBA

Just below Orion the Hunter and just to the west of Canis Major the Greater Dog is Lepus the Hare. Lepus gets little press compared to Canis Major and Orion, but, surprisingly, it is relatively bright and easy to see even in light-polluted skies. Lepus is composed of eight main stars, the four brightest of which look like a squished rectangle south of Rigel in Orion. Lepus hardly resembles the animal it is supposed to represent.

Halfway between Lepus and Canopus, the second-brightest star, is the loose collection of stars Columba the Dove. It requires a somewhat darker sky to find due to its relative dimness and low position in our Northern Hemisphere sky. Unlike Lepus, which comes down to us from the ancients, Columba is a more "modern" constellation that is sometimes attributed to the Dutch astronomer Petrus Plancius (1552–1622) and appeared in Johann Bayer's famous sky atlas *Uranometria* in 1603. Lepus may have been a hare being pursued by Orion and his hunting dogs, while Columba was created to fill a bit of space in the southern sky south of Lepus. It is hard to picture Columba as a dove with a sprig in its mouth. Both Lepus and Columba have telescopic objects of great interest to astronomers.

LYNX

Lynx the Lynx is directly north of Cancer the Crab. Lynx is a "modern" constellation introduced by the famous Polish astronomer Johannes Hevelius (1611–1687), who wanted to fill a gap in the sky between Ursa Major the Great Bear and Auriga the Charioteer. Lynx is a zigzag of faint stars that looks nothing like its ferocious namesake. It does have several telescopic objects of interest to astronomers and is a good challenge for constellation observers.

LYRA

A lyre is a stringed musical instrument like a harp. Lyra the Lyre represents the golden lyre given to Orpheus by Apollo, the god of music. Orpheus was a legendary musician, poet, and prophet in ancient Greece. Lyra, although small, upholds the myth of a golden lyre well, because it is an easily recognizable constellation. The main portion of Lyra is a small parallelogram of four bright stars.

Just above (north of) the parallelogram is Vega, the fifth-brightest star (figures 4.9a and 4.9b). Vega is the brightest star in the Summer Triangle of Altair, Deneb, and Vega.

MONOCEROS

Monoceros the Unicorn is a boring collection of faint stars that has spectacular telescopic objects that are favorites of amateur and professional astronomers. Monoceros lies south of Canis Minor the Lesser Dog and east of Orion the Hunter.

OMEGA CENTAURI

Omega Centauri is not a constellation. It is an object in Centaurus the Centaur, a mythical creature with the head, arms, and torso of a man and the body and legs of a horse. Centaurus is a large constellation, which, unfortunately, is too far south for its two brightest stars Alpha and Beta Centauri to be visible from the continental United States except for the southern tips of Florida and Texas.

Omega Centauri is the largest globular cluster in the Milky Way. A globular cluster is a roughly spherical cluster of hundreds of thousands of stars held together by gravity. The Milky Way has approximately 150 globular clusters, the biggest and brightest of which is Omega Centauri. Unfortunately, it is very low in the southern sky, but farther north than the stars Alpha and Beta Centauri, and it is visible in the southern part of the continental United States. In southern Arizona it is visible to the naked eye in a dark sky free of the Moon (figure 4.11b). It is readily visible through binoculars even in moderate light pollution. In the Southern Hemisphere Omega Centauri is a truly magnificent sight in a dark sky.

OPHIUCHUS

Directly to the north of Sagittarius the Archer and Scorpius the Scorpion is the large constellation Ophiuchus the Serpent Holder (figure 4.6). Ophiuchus gets less press than it deserves, probably because of its tongue-twisting name.

Ophiuchus is supposed to be a man holding a large snake represented by Serpens Caput (the head) on its western end and Serpens Cauda (the tail) on its eastern end. Ophiuchus the constellation is a misshapen oval of about ten brighter stars with lines of stars (the Serpens) on either end. Ophiuchus is most deserving of our attention because it and Serpens Cauda are contiguous to the western edge of the Milky Way.

Ophiuchus has reasonably bright stars and contains many delightful binocular and telescopic objects favored for viewing and study by astronomers. It is one of the thirteen constellations of the zodiac.

ORION

Orion the Hunter is a large, very bright constellation (figures 4.3b, 4.4, 4.8, and 6.3b). It has more bright stars than any other constellation and is probably the second most widely known and recognized constellation after the Big Dipper (Ursa Major).

Orion dates from prehistoric times and is associated with many legends. In one legend, a great hunter, Orion, boasted he was invincible just like the gods. This angered the deities, and Orion was killed by the sting of a scorpion now known as Scorpius. Then the gods felt sorry and placed Orion in the sky along with Scorpius, but they kept the two far apart. Scorpius does not fully rise until Orion has set. Perhaps Scorpius is chasing Orion in the sky, or Orion is chasing Scorpius looking for his revenge.

Orion is called a winter constellation, because it rises in the early evening in the late fall and early winter. However, Orion is visible for a good part of the year if you are willing to look for it very early in the morning before sunrise in the late summer.

Orion lies in a vast star-forming region of the Milky Way. Even though the closest stars in Orion are hundreds of light years away, they are bright because they are supergiants giving off enormous amounts of energy. They have interesting names and contrasting colors. All of them are far larger and brighter than the Sun, and they are destined to die "soon," astronomically speaking.

Somewhat dimmer than the main stars in Orion are the stars in the sheath of Orion's sword hanging from his belt. This area contains the Great Orion Nebula (M42 on Charles Messier's catalog of important sky objects), a stellar nursery where new stars are being born. While it is easily visible to the naked eye as a fuzzy region, it shows itself to be a complex mixture of gas, dust, and stars through a small telescope.

PEGASUS AND ANDROMEDA

Pegasus the Winged Horse and Andromeda the Maiden are really one large constellation (figures 4.9b, 5.2a, and 5.2b). Andromeda was chained to a rock as a sacrifice for a sea monster (Cetus). Fortunately, she was rescued by Perseus the Hero, who lies close by in the sky. Pegasus looks nothing like a horse. It is a large square of relatively bright stars. Inside the square are dimmer stars, and the number of those you can see on a given night indicates how dark the sky is. Andromeda also doesn't look anything like a maiden. Mainly it consists of three stars extending northeast from the square of Pegasus.

The northeastern star in the "square" of Pegasus is Alpheratz, which is technically in Andromeda, though I like to think of it as completing the square of Pegasus. Extending from Alpheratz to the northeast are two other bright stars, Mirach and Almach, in that order, that together make up most of Andromeda. Seven degrees north of Mirach is a fuzzy patch, which is the Andromeda Galaxy, also known as M31 because the French astronomer Charles Messier included it as the thirty-first entry in his catalog of important sky objects (figure 5.2c). The Andromeda Galaxy is approximately 2.5

light years away, and it is one of the most distant objects visible to the naked eye. It can be seen with the unaided eye in a dark sky, and it is a wonderful sight through binoculars or a small telescope. M31 is a large spiral galaxy comparable to our parent galaxy, the Milky Way. In fact, the Andromeda Galaxy and the Milky Way are the largest galaxies in the Local Group of galaxies of approximately forty-plus members traveling together through space.

PERSEUS

Perseus the Hero rescued the maiden Andromeda, who was chained to a rock, from the sea monster Cetus. Perseus looks like a stick-man with two very long legs (figures 4.3b and 5.6). It is sort of an upside-down *V*. Perseus has bright stars and an easily recognized configuration. In the eastern limb of Perseus is its brightest star, Mirfak, which lies 593 light years away and has a radius fifty-seven times that of the Sun. In the western limb of Perseus is the star Algol ("demon" in Arabic).

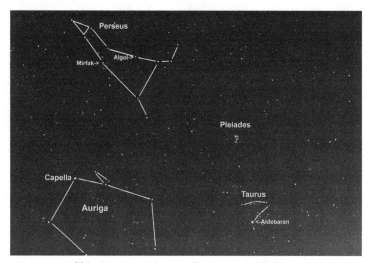

FIGURE 5.6 Telephoto view showing Perseus with Mirfak and Algol, part of Auriga with Capella, the Pleiades, and the main portion of Taurus with Aldebaran.

DOUBLE CLUSTER

While Perseus and Cassiopeia are bright constellations with much to see in their main portions, there is a wonderful sight halfway between Perseus and Cassiopeia, the Double Cluster. Two bright star clusters lie close to each other and are visible to the naked eye even in moderate light pollution. These clusters are technically in Perseus and are often listed by their New General Catalogue (NGC) designations, NGC 869 (to the west) and NGC 884 (to the east; figure 5.7). The Double Cluster is gorgeous through binoculars or a small telescope. It is a must-see object and should become a familiar friend. The clusters are thought to lie 7,600 light years away and seem to be traveling through space together.

FIGURE 5.7 Telescope view of the Double Cluster in Perseus.

PISCES

Pisces the Fishes is one of those constellations everyone has heard of but few have actually seen. Pisces is a zodiacal constellation. Like several other zodiacal constellations, it has a hard time living up to its reputation, since it is faint and somewhat hard to find, though it is large.

Pisces is supposed to be two fish, one facing west below Pegasus the Winged Horse and the other facing north, lying east of Pegasus and west of Aries the Ram (figure 5.2b). Naturally, Pisces looks nothing like fish. To see Pisces, find the bright square of Pegasus and look south to see seven fainter stars that form a misshapen heptagon, a seven-sided polygon. Six faint stars extend east of the heptagon, and then five stars extend north between Pegasus on the right (west) and Aries on the left (east), to complete the constellation. It's a tough constellation to identify, but it gives you good feelings of accomplishment.

PLEIADES

The Pleiades, or Seven Sisters, is a small cluster supposedly made up of seven stars, but most people only see six stars. Technically, the Pleiades are part of Taurus the Bull, but they are frequently talked about as if they were a separate constellation (figures 4.3a, 4.3b, 5.6, and 5.8). Charles Messier listed it as the forty-fifth object in his catalog, so it is also known as M45.

If you have a good dark sky with a steady atmosphere, you can see seven or more stars in the Pleiades. In fact, the Pleiades contains hundreds of stars. We only see the brightest members of the cluster with our naked eyes. Most of the time, I see six stars in the Pleiades. On a few rare nights, I have seen eight or nine. The ancients who described the Pleiades probably saw no more stars than we see in it, though they did have darker skies. They probably picked seven for the number of the sisters because that seems a more magical number than six.

FIGURE 5.8 Telescope view of the Pleiades star cluster, the Seven Sisters. The Pleiades is part of Taurus the Bull and consists of large, hot, white, or blue-white stars surrounded by nebulosity.

Look at the Pleiades with a pair of binoculars or a small telescope. You will be astounded by the number of stars of varying brightness scattered among the six bright stars in the cluster. The Pleiades is relatively young—only about one hundred million years old. It lies 440 light years from us, and its members are hot blue stars, though they look mainly white to my eyes. The stars are surrounded by nebulosity, which can sometimes be seen with a very good telescope but is mostly noted on long-exposure photographs (figure 5.8).

SAGITTARIUS

Sagittarius the Archer is traditionally said to represent a centaur drawing a bow. A centaur is a half-man (or woman), half-horse mythical figure with the four legs of the horse and the upper body of a man or woman with arms. I prefer to think of Sagittarius as a teapot, because its bright stars have a configuration that closely resembles a teapot (figure 5.9). It has eight bright stars, with its spout composed of three stars to the west and its handle composed of four stars to the east. Its top points north and is a triangle composed of the top stars of the spout and handle plus a single star at its apex.

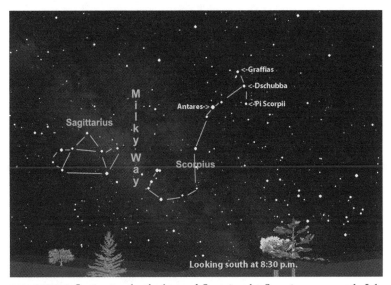

FIGURE 5.9 Sagittarius the Archer and Scorpius the Scorpion on an early July evening. The very center of the Milky Way is in Sagittarius.

Sagittarius really comes into its own during the dark of the Moon. Get away from city lights and look at Sagittarius under a dark sky when it is well above the horizon. You will be stunned by the glorious summer Milky Way that runs right through Sagittarius. The Milky Way is the official name of our own Galaxy (often capitalized if referring to the Milky Way), which contains the Sun, the solar system, and hundreds of billions of other stars. In the Southern Hemisphere, Sagittarius is much higher in the sky, and the Milky Way is even more spectacular.

The very center of the Milky Way is in Sagittarius, and it contains a black hole with a mass of approximately three to four million times that of the Sun. Black holes are strange objects with so much mass concentrated into such a small relative volume that not even light can escape from them. Anything falling into a black hole can never escape. The effects of the Milky Way's black hole can be seen with radio telescopes as it goes about gobbling up nearby gas, dust, and stars. Fortunately, it is way too far away and way too small compared with the size and mass of the rest of the Milky Way to concern us.

SCORPIUS

Scorpius the Scorpion has it all—bright stars and a fascinating myth (figure 5.9). Scorpius in some tales is the giant scorpion that killed Orion the Hunter, who boasted he feared no animal. Scorpius and Orion are never in the sky at the same time. Orion sets in the west as Scorpius rises in the east. Is the scorpion chasing Orion out of the sky, or is the scorpion fleeing Orion, who is looking for revenge?

Of all the constellations representing a legendary beast or hero, Scorpius most clearly resembles what it is supposed to represent, from its head and heart to its long, curved tail with a stinger. Its pincers are toward the west and its curved tail toward the east. A portion of the Milky Way goes through Scorpius, and the giant red star Antares is at the heart of the scorpion. Every month the Moon passes through Scorpius close to Antares on its monthly circuit of the sky. The stars Graffias, Dschubba, and Pi Scorpii, from east to west, respectively, make up the head of the scorpion (figures 4.6 and 5.9).

TAURUS

Taurus the Bull includes the Pleiades, but the main part of the constellation looks like a large V lying on its side. The point of the V is toward the south and the open part of the V is toward the north. The brightest star in Taurus is orange/red Aldebaran (figures 4.3a, 4.3b, 4.4, 5.6, and 6.3b). If you look at Taurus with binoculars, you will be amazed at the number of dimmer stars present in Taurus. The stars of Taurus are part of the Hyades star cluster. These stars are going through space together and are 150 light years away from us. The Hyades cluster contains approximately three hundred to four hundred stars.

Aldebaran has a distinctive red or orange color that contrasts with the white or bluish-white color of the stars of the Hyades. Star colors as we see them in most cases are subtle, and they are easily influenced by the sky conditions, the apparent brightness of the star to our eyes, and whether we observe them with some type of optical aid, such as binoculars.

Aldebaran is not part of the cluster, but it adds a sparkle to it. It looks bright in comparison to the stars of the Hyades, because it is indeed a large bright star, and it is much closer to us.

VIRGO

Virgo the Virgin is like an elongated, somewhat bent rectangle of stars with an extension of stars to the west. It is a large constellation, but only modestly bright. The ecliptic, the path the Sun follows in the sky, goes through Virgo. Much of Virgo's interest to amateur and professional astronomers is the large number of important telescopic objects, mainly galaxies, spread through the constellation. The Virgo Galaxy Cluster is a cluster of up to two thousand galaxies approximately fifty-four million light years away. These galaxies are centered in Virgo and spread out over an approximate 8-degree extent. Many of the larger brighter members of the cluster are readily visible through amateur telescopes, making Virgo an observing favorite in the spring and summer skies. Of course, the stars in Virgo have nothing to do with the galaxies, being only nearby foreground objects sitting in front of distant galaxies.

The brightest star in Virgo is Spica, the sixteenth-brightest star (figure 4.7). When Virgo is high above the horizon and the night is dark and steady, Spica has a remarkable blue color. Much of its radiation is ultraviolet light rather than visible light.

6

Other Wonders of the Night

EARTH'S SHADOW AND THE BELT OF VENUS

IT IS dark at night because we are in Earth's shadow, which hides the Sun. Near sunrise and sunset, it is possible to see Earth's shadow above the horizon (figure 6.1). This is a twice-daily event that is visible for several minutes. No doubt most of us have seen it many times but have not recognized it for what it is. In fact, many gorgeous pictures taken near sunrise and sunset show Earth's shadow without the photographers realizing it.

As the Sun sets in the west, a dark blue band, Earth's shadow, rises in the east. This band is bounded above by a pinkish or orange glow called the antitwilight arch. This arch is also called the Belt of Venus, and its color is caused by backscattering of red light from the setting Sun. Earth's shadow and its bordering Belt of Venus stretch nearly 180 degrees across the eastern horizon.

Just before sunrise a similar phenomenon occurs. In the western sky you can see a dark blue band slowly sink into the horizon. Above it is the reddish antitwilight arch. These effects are best seen during a cloudless twilight before sunrise or just after sunset. On some days it is more distinct than on others, but once you have seen it, you will be hooked on it and look for it anytime you are outside near sunrise or sunset.

FIGURE 6.1 Telephoto view of the eastern horizon just after sunset. The dark band along the horizon is Earth's shadow. This black-and-white image does not show the Belt of Venus, a red or orange band along the top of Earth's shadow. Just above Earth's shadow is the rising, nearly full Moon.

ZODIACAL LIGHT

A fun observing challenge is to look for the **zodiacal light**. From the somewhat lower latitude of Tucson, Arizona, you can often see it throughout the year. At midlatitudes it is best seen in the west after evening twilight in the spring, or in the east just prior to morning twilight in the autumn. The zodiacal light is a very large pyramidal glow in the western sky after sunset or in the eastern sky before sunrise. It is sometimes called the false dawn if visible prior to morning twilight. The zodiacal light is surprisingly bright and extends at least halfway up the sky from the horizon (figure 6.2).

The first time you see the zodiacal light, you might mistake it for the glow of city lights. It is the glow of microscopic dust particles left by comets and asteroids. These innumerable fine grains lie along the orbital plane of the solar system. The Sun's light reflects

FIGURE 6.2 Wide-angle view of the western horizon after sunset in the late summer. The zodiacal light is quite bright (arrow). To the right (north) of the zodiacal light is the setting Milky Way.

off of them, producing a distinct glow best seen during those times when the plane of the solar system (the ecliptic) is nearly perpendicular to the eastern or western horizon.

OUR GALAXY

The Milky Way is the name for our parent galaxy. In astronomical journals and books, the Milky Way is sometimes called the Galaxy with a capital *G*. It is a very large spiral galaxy with several spiral arms. We are in one of its spiral arms approximately twenty-five thousand light years from the center of the Galaxy, which is in the constellation Sagittarius. At the very center of the Galaxy is an enormous black hole that is gobbling up dust and gas at a ferocious rate. It is obscured by the dust and gas and is best studied using radio telescopes and infrared instruments. Fortunately, it is way too far away to bother us.

From our point of view, the Milky Way wraps completely around the sky, and there is a "summer Milky Way" and a "winter Milky

Way" (figures 6.3a and 6.3b). The summer Milky Way runs from Cassiopeia through Cepheus, Cygnus, Aquila, contiguous to the edge of Serpens Cauda and Ophiuchus, and ends up in Sagittarius. For Northern Hemisphere observers, the summer Milky Way is brighter and more spectacular than the winter Milky Way. We are looking more toward the center of the Galaxy when viewing the summer Milky Way and away from the center of the Galaxy when viewing the winter Milky Way.

The winter Milky Way stretches from Cassiopeia and Perseus right through Auriga and across the bottom of Gemini. It lies somewhat to the east of Orion and Canis Major and then trails off to end up in the dim constellation Puppis the Stern (of a celestial ship). On a cold winter's evening, it is nice to see the subtle but apparent Milky Way contrasted with bright Orion and Canis Major.

Spiral galaxies like the Milky Way are really flattened disks when seen on edge. The Milky Way with its spiral arms contains hundreds of billions (!) of stars. Its total mass is probably over a trillion times the mass of the Sun. When we look in the sky toward what we call the Milky Way, we are looking at our parent Galaxy on edge. We see thousands of stars seemingly piled on top of one another, thus giving the illusion of milk spilled ("*via lactea*" in Latin) across the sky. When we look away from the plane of the Milky Way's disk, we see many fewer stars, and there is no "milky" band. The term "Milky Way" means both that band of stars, gas, and dust stretching across the sky in the plane of our Galaxy, and it is also the general name applied to our Galaxy.

All of the stars scattered around the sky separate from the Milky Way band, nevertheless, are really contained in the Milky Way, our parent galaxy. Most other galaxies contain billions of stars, but we cannot see individual stars in them, as they are way too far away. Only the largest professional telescopes can view individual stars in nearby galaxies. When you see a picture of another galaxy, you should realize the stars scattered in front of the galaxy or around it are foreground stars in the Milky Way and are far closer to us than the galaxy.

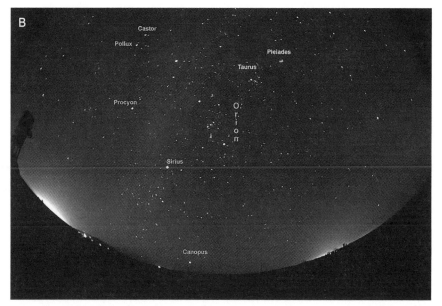

FIGURE 6.3 (a) The summer Milky Way setting on a late August evening. (b) The winter Milky Way as seen from southern Arizona. It runs through Auriga and east of Orion and Canis Major. Sirius, the brightest star, and Canopus, the second-brightest star, are conveniently situated near each other, not far from the winter Milky Way. Of course, all stars we see are contained in the Milky Way.

The overhead summer Milky Way at a dark-sky location is an astronomical sight not to be missed. The winter Milky Way from a dark-sky location is also simply magnificent. Scan through the Milky Way with binoculars to see countless stars, many star clusters, and areas of nebulosity. The Milky Way is so spectacular that viewing it rivals other top astronomical experiences. The Milky Way is perhaps not quite as stunning as a total solar eclipse, but it has the advantage of being present most of the time if the Moon is out of the way, the sky is clear, and there is little light pollution.

There are thousands of star clusters and nebulae throughout the Milky Way. Most of them are telescopic objects, the largest and brightest of which are suitable for amateur astronomers to observe or photograph. All of the larger, brighter star clusters and nebulae are objects of study by professional astronomers. Several of them are even bright enough to be seen with the naked eye in a dark sky and have been known since ancient times, though their composition was not understood until more modern times.

The Great Orion Nebula (M42), the Hyades, the Beehive (M44), the Pleiades (M45), the Double Cluster, and Omega Centauri are especially interesting sights and bright enough to be visible to the unaided eye through moderate light pollution. These objects have already been discussed along with their parent constellations in chapter 5, but one should understand they are top attractions in the Milky Way.

METEORS, METEOR SHOWERS, AND COMETS

SHOOTING STARS

All night long, an occasional **shooting star** or **falling star** streaks across the sky. Some even leave glowing trails for a while. Stars, of course, don't fall out of the sky. The scientific name for a shooting star is **meteor**. Aristotle coined the word "meteor" after the Greek word for "things in the air," because he believed meteors were entirely an Earth-based phenomenon. We now know they come from outer space. Earth is constantly bombarded with debris (meteoroids) from space. Much of this debris represents bits of rock and metal left over from the formation of the solar system. Also, when comets swing by the Sun, they leave debris in their wake— tiny rocks and ices—that Earth frequently encounters.

COMETS

Comets are large "dirty snowballs" that consist of various ices of frozen water, carbon monoxide, carbon dioxide, ammonia, and methane, mixed with dust and rocky debris. This icy, dusty mixture is the "nucleus" of the comet, which is stable in the cold outer reaches of the solar system. When a comet travels into the inner

FIGURE 6.4 Comet NEOWISE rising above the Catalina Mountains north of Tucson, Arizona, on the morning of July 9, 2020.

portion of the solar system to the Earth's orbit or farther, the nucleus is heated by the Sun and gives off large volumes of thin gas as its ices vaporize.

Most comet nuclei are only a few miles across, although a few large comet nuclei may be fifteen to thirty miles in size. Even so, as the nucleus is heated by the Sun, a very large cloud of gas, the coma, surrounds the relatively tiny nucleus, and one or more long tails of gas, dust, and ionized materials may stream out from the comet nucleus. A comet coma is often hundreds to thousands of miles in diameter. A comet tail may stretch for millions of miles. Naturally, the coma and comet tail, though quite large, have very low, almost vacuum-like, densities. Most comets are faint telescopic objects with a small coma and little or no tail. Every few years, a comet appears that is readily visible to the naked eye, with a long tail, putting on a spectacular show for several days to weeks (figure 6.4).

There are a few hundred comets that are known as periodic or short-period comets, which have orbital periods of two hundred years or less. Most of these have been observed more than once, and their return to the inner solar system is predictable. The most famous short-period comet is Comet Halley, which returns every seventy-five to seventy-six years. It has been observed since at least 240 BC. Edmond Halley (1656–1742) first noted that there was a bright comet that returned every seventy-five to seventy-six years. In 1705, he predicted its return in 1758, even though he knew he would not be alive to see its return. It was spotted on December 25, 1758, confirming Halley's prediction, and the comet was later named after him.

Unlike Comet Halley, most periodic comets are faint and uninspiring to the casual observer. In fact, most comets, whether periodic or from the far reaches of the solar system with periods of thousands of years, are faint, telescopic objects. Those rare comets that happen to be especially bright come from the outer reaches of the solar system, have not been observed previously in our historical time frame, and have periods of thousands of years, never to return in one's lifetime. The rareness of bright comets coupled with their seemingly sudden appearance in times before telescopes led ancient peoples to consider comets harbingers of doom. By Halley's time, it was realized comets were natural celestial phenomena far outside Earth's atmosphere. It was soon determined that they follow the laws of planetary motion, as first enumerated by Isaac Newton (1642–1727) in 1687.

METEOR SHOWERS

Rocky, dusty, and icy debris is scattered throughout the solar system, leftover from the formation of the planets and their moons. The main portion of this material lies in the asteroid belt between Mars and Jupiter. Some of it collected to form icy comet nuclei in the far outer reaches of the solar system. Some of it lies between the planets. Even though most of the solar system is empty space, there are enough scattered tiny bits of debris that thousands of tons of material hits Earth every year. This seems like a lot, but it is trivial compared to the large size of Earth.

A meteor streaking across the sky looks like a falling star, but the real stars remain firmly in place. Meteors are typically the size of a grain of sand. When they pass through Earth's atmosphere at high speeds, they heat up, glow brightly, and finally vaporize. Very bright meteors are called "fireballs." An exceptionally large meteor that survives its passage through the atmosphere and ends up on the ground is called a **meteorite**.

As Earth moves around the Sun, it occasionally encounters a rich lode of debris when it crosses the path of a (usually periodic) comet, and a meteor shower occurs. Comets are constantly dropping bits of dust, rock, and ice along their paths. There are more than ten well-known meteor showers that are potentially visible every year. Few of these produce many bright meteors, and all meteor showers are unpredictable, unfortunately often not living up to their hype. Cloudy weather or a bright Moon can ruin what would otherwise be a good event. Even so, there are three reasonably predictably good meteor showers most years—the Perseids in August, the Leonids in November, and the Geminids in December.

During a meteor shower, most of the meteors appear to come from a point in the sky. This point or "radiant" is in the constellation for which the shower is named.

THE PERSEIDS

The Perseids meteors are from Comet Swift–Tuttle, a periodic comet with an irregular orbital period of 133 years. It was discovered independently by Lewis Swift (1820–1913) on July 16, 1862, and Horace Parnell Tuttle (1837–1923) on July 19, 1862. The radiant for the Perseids is in the constellation Perseus the Hero, who rescued the maiden Andromeda from the sea monster Cetus. The Perseids are active for several days in mid-August. The peak of the shower is often predicted to be on the night of August 11 and the morning of August 12, though that can vary from year to year. The Perseids are sometimes referred to as the "tears of Saint Lawrence," returning to Earth every year around August 10, the date of the saint's martyrdom in 258 AD in which he was said to have been burned alive on a gridiron!

Even though Perseids radiate from Perseus, they can be seen all over the sky, and it is not necessary to look at Perseus to see lots of meteors. If you have good weather and a dark sky you can see, on average, a meteor a minute when the shower is going full blast. Perseus is well above the horizon by 1:00 a.m. in mid-August. From then until morning twilight, the meteors should get better and better. While you are at it, look at the nearby Pleiades, Taurus, and Auriga. For all meteor showers, get out a good lawn chair, place it in a convenient location, bundle up (or put on insect repellent in warm weather), and lean back and enjoy the show with hot chocolate, coffee, tea, or another beverage (nonalcoholic to preserve your night vision!).

THE LEONIDS

The Leonid meteors are from bits of debris from Comet 55P/Tempel–Tuttle. The comet was discovered independently by Ernst Tempel (1821–1889) on December 19, 1865, and Horace Parnell Tuttle on January 6, 1866. This comet is a periodic comet—hence the *P* in its name—and it returns to the inner solar system every thirty-three years. Its last closest approach was in 1998, and its next close approach will be in 2031. It is a not a particularly large or bright comet, but its debris gives us a generally good meteor shower every year, and a spectacular meteor shower from time to time.

Typically, the Leonids peak on the night of November 17 and the morning of November 18, but, like other meteor showers, the peak can vary from year to year. The Leonids are better after midnight and generally will get better until bright morning twilight. The front part of Leo is just above the eastern horizon at 1:00 a.m., and the entire constellation is up by 2:00 a.m. Even though the

Leonids radiate from Leo, they can be seen all over the sky, and it is not necessary to look at Leo to see lots of meteors. Sometimes it is more convenient to orient your chair away from the east to keep Leo and a neighbor's irritating light or the Moon toward your back and out of your eyes.

THE GEMINIDS

The radiant for the Geminids is Gemini the Twins. The Geminids typically have their peak on the night of December 13 and the morning of December 14, though this can vary from year to year. It is too bad the Geminids do not take place at a warmer time of year—probably one of the reasons why the Geminids do not get as much attention as the Leonids or the Perseids. The Geminids may be the most consistently good meteor shower from year to year. It is said to produce up to 120 bright, multicolored meteors an hour at its peak, but do not expect to see this many. The Geminids do produce a lot of bright meteors and are well worth bundling up to observe on a cold December evening.

The Geminids are produced by debris left behind by the asteroid 3200 Phaeton (also spelled Phaethon), which was discovered in 1983. The "3200" means Phaeton was the 3,200th asteroid to have its orbit precisely known. Phaeton is named after Phaethon, the son of the sun god Helios.

Phaeton comes closer to the Sun than any other named asteroid. It has a dust tail and has been seen ejecting dust. Shortly after Phaeton was discovered, the famous astronomer Fred Whipple (1906–2004) noted that the orbit of Phaeton was similar to the orbit of Geminid meteors. Whipple developed the modern theory for comet composition known as the "dirty snowball" hypothesis, in which comets consist of various frozen "ices" mixed with dust. Phaeton is likely the rocky skeleton of a comet that lost most of its outer covering of ice.

Even though the Geminids radiate from Gemini, they can be seen all over the sky, and it is not necessary to look at Gemini to see lots of meteors. Gemini rises high enough to begin viewing it by 9:00 p.m., and it stays up all night.

Gemini is overhead at 1:30 a.m. However, it's hard to look directly overhead for any length of time. I find it is most comfortable to look at the sky about halfway up from the horizon. This is easily done while sitting in a favorite lawn chair. You can look in any direction that strikes your fancy, since the meteors will appear to radiate from Gemini and travel to all parts of the sky. If you choose to look toward the south for a while, you will enjoy mighty Orion the Hunter with his faithful hunting dog Canis Major following in his footsteps. This will give you the opportunity to see Sirius, the brightest star. If you live far enough south, you can even see Canopus, the second-brightest star.

7

The Seasons and the Calendar

THE SEASONS

EARTH IS like a giant gyroscope, completing one spin around its axis each day. As Earth moves around the Sun, its axis keeps pointing to the same place in the sky, but the tilt of Earth toward the Sun constantly changes. This gives rise to our seasons. The North Pole is tilted toward the Sun for half of the year and tilted away from the Sun for the other half (figure 7.1). This changes the amount of sunlight a given place receives throughout the year. The Northern Hemisphere leans toward the Sun during summer and leans away from the Sun during winter. In the spring and fall, both hemispheres have about the same tilt toward the Sun.

VERNAL (MARCH OR SPRING) EQUINOX; AUTUMNAL (SEPTEMBER OR FALL) EQUINOX

Because Earth makes one complete orbit around the Sun in a year, the Sun appears to move slowly across the sky from day to day. When the Sun crosses directly over the equator from the southern to the northern part of the sky, winter officially gives way to spring. This is the vernal (March or spring) equinox. It happens around March 21 every year. When the Sun appears to cross the celestial

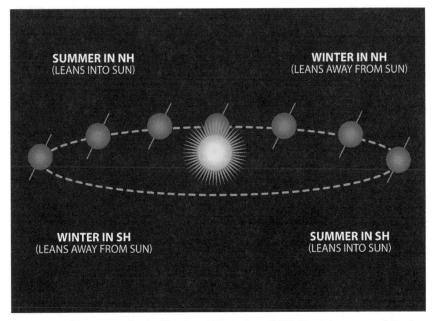

SUMMER IN NH
(LEANS INTO SUN)

WINTER IN NH
(LEANS AWAY FROM SUN)

WINTER IN SH
(LEANS AWAY FROM SUN)

SUMMER IN SH
(LEANS INTO SUN)

FIGURE 7.1 The seasons. In the Northern Hemisphere (NH), Earth leans toward the Sun in the summer and away from the Sun in the winter. In the Southern Hemisphere (SH), Earth leans in the opposite direction. Drawing by Debra Bowles.

equator from the northern part of the sky to the southern part of the sky, summer gives way to autumn at the autumnal (September or fall) equinox, which takes place around September 21 every year.

"Equinox" means "equal night"—the hours of night and day are equal. However, that is not quite correct. The hours of day and night are close to, but not exactly, equal at these times. For the latitude of Tucson, Arizona, approximately 32 degrees north, the times of equal daylight (from sunrise to sunset) and night (from sunset to sunrise) happen five to six days prior to March 21 and five to six days after September 21. Even so, there is somewhat more daylight than night, because the atmosphere acts as a giant lens and bends the Sun's light up above the horizon for a few minutes before sunrise and after sunset. However, on the vernal and autumnal equinoxes, the Sun rises due east and sets due west.

SUMMER (JUNE) SOLSTICE;
WINTER (DECEMBER) SOLSTICE

In various parts of Europe, time centered on the summer solstice was known as midsummer. Preindustrial agricultural societies followed the seasons much more closely than we do, because their crop planting and harvesting and their very existence depended on the seasons. They used the sky for their timekeeping and calendar reckoning. The Sun's position in the sky was accurately noted by ancient observers, who closely studied the sky just before sunrise and just after sunset, which gave them a good estimate of where the Sun was with respect to the starry background. Technically speaking, all of the stars we see in the night sky are moving with respect to the solar system, but they are so far away that their movements appear very tiny and take many years or even decades to detect with sophisticated instruments.

The solstices are those times at which the Sun is at its greatest distance north (summer solstice) or south (winter solstice). The summer (June) solstice takes place around June 21, and the winter (December) solstice takes place around December 21. In our part of the world, the shortest night of the year is on the summer solstice, and the longest night of the year is on the winter solstice, though the lengths of daylight a few days before and after the solstices are approximately the same. The solstices are reversed in the Southern Hemisphere. Our day is defined for civil purposes as being twenty-four hours long, with each day having the same length as any other day. When we use the term "day" with respect to the solstices or equinoxes, we mean the time of daylight from when the Sun rises until it sets.

The exact date of a solstice or an equinox varies by a few days every year, because our calendar has 365 days in a year (or 366 days in a leap year), but Earth takes 365.256 days to complete one orbit around the Sun. This, coupled with small variations in the daily rotation of Earth and the slow wobble of Earth's axis, causes the solstice dates to vary. Most of us hardly give the solstices a glance, but they had more importance to preindustrial agricultural societies. The solstices remind us to salute our forefathers. After all, without them, we wouldn't be here to appreciate what the heavens have to offer.

THE TROPICS

The Tropic of Cancer is at 23.44 degrees north latitude. Also known as the Northern Tropic, it is the north circle of latitude at which the Sun appears directly overhead at its most northern point at the summer solstice. The equivalent line of latitude south of the equator, at the winter solstice, is called the Tropic of Capricorn at 23.44 degrees south. The region between these two lines is the **tropics**. The term "tropic" is derived from a Greek word meaning turn, change of direction, or change of circumstance, because the Sun appears to turn back at the solstices and reverse its direction in the sky.

The tropic names come to us from the last centuries BC when the Sun was in the constellation Cancer the Crab at the summer solstice and in Capricornus the Sea Goat at the winter solstice. Since Earth's axis slowly moves through a circle (*precession of the axis*) in a cycle of approximately twenty-six thousand years, the Sun is now in the constellation of Taurus the Bull at the summer solstice and in Sagittarius the Archer at the winter solstice.

THE CALENDAR

Leap year is a special day for me, as it is my wedding anniversary. I once told my wife we only had to celebrate our anniversary once every four years. However, I quickly realized it was in my physical and fiscal best interest to celebrate our anniversary every year, typically on March 1. As mentioned before, our civil calendar is based on the Sun and is defined to have twenty-four hours in a day and 365 days in a year. The orbital period of Earth, upon which our year is defined, is, unfortunately, not exactly 365 days. It is currently 365.256 days. Over hundreds of years, the starting dates of the seasons (the solstices and the equinoxes) would drift if we simply defined our year as 365 days.

To address this problem, Julius Caesar introduced the Julian calendar in the Roman Empire in 46 BC. This calendar set a standard year to be 365 days and every four years added a leap year of 366 days, with the extra day added to the end of February. The year started on January 1. The Julian calendar works well, but for every thousand years, it gives an error of eight days. By the sixteenth century, it was apparent that calendar reform was again necessary.

Our present calendar was introduced by Pope Gregory XIII in 1582. In the Julian calendar, all century years (1600, 1700, et cetera) were leap years. However, in the Gregorian calendar only those century years divisible by four hundred are leap years. For example, 2000 is a leap year, but 1900 and 2100 are not. The Gregorian calendar is accurate to about one day in three thousand years. Britain and the American colonies did not adopt the Gregorian calendar until 1752, when the date jumped from September 2 to September 14 overnight!

LUNAR CALENDARS

Our civil calendar is based on the Sun with a day defined as twenty-four hours, and a year defined as 365 days (366 days in a leap year), as discussed above. While our civil calendar is more than two thousand years old and is based on the system devised under Julius Caesar, it is relatively new compared to lunar calendars that are based on the phases of the Moon. Lunar months average 29½ days (a lunar cycle from new Moon to new Moon or full Moon to full Moon). The historic Chinese calendar starts each month on the new Moon, while the Hindu calendar begins each month on the full Moon. The Hebrew calendar and the Islamic calendars are based on the first sighting of a very thin lunar crescent after new Moon.

Because lunar calendars do not coincide with our everyday civil calendar, they are now used mainly for religious purposes.

COORDINATED UNIVERSAL TIME (UTC)

UTC stands for **coordinated universal time**, which is the official time standard for the world. It is roughly equivalent to **Greenwich mean time** (GMT), the time at Greenwich, England (0 degree longitude). UTC and GMT do not have daylight saving and they use a twenty-four-hour clock notation. They are seven hours ahead of mountain standard time (MST), the local time for Tucson, Arizona. In the United States, times listed for an event of national interest are usually given in eastern standard time (EST) or eastern daylight time (EDT), and you must mentally calculate the event's time for your time zone.

The times listed for astronomical events are most often given in terms of UTC so astronomers around the world speak the same language with respect to time. To convert from UTC to MST, for example, subtract seven hours using twenty-four-hour notation. Thus, 1:54 UTC (early morning at Greenwich) is 18:54 (6:54 p.m. MST) in Tucson on the preceding day. For eastern standard time this would represent 20:54 (8:54 p.m. EST), or 21:54 (9:54 p.m. EDT) on the preceding day.

SELECTED CALENDAR DATES

The Sky Spy column in the *Arizona Daily Star* runs in the *Caliente* section of the newspaper every Thursday. Because interesting days sometimes fell on Thursday, I often wrote about the astronomical significance of that date, if any. Every year, Thanksgiving falls on the fourth Thursday in November in the United States, so I would always have a Thanksgiving message for that column.

Sometimes, Christmas or New Year's falls on Thursday, and I would try to give an astronomical perspective on that day. Other national holidays in the United States, such as Presidents' Day, Martin Luther King Day, Memorial Day, and Labor Day always fall on a Monday and usually are not covered in the Sky Spy column. July 4, Independence Day, and November 11, Veterans Day, in the United States sometimes do fall on Thursday, and I have dedicated columns to them. There are other, somewhat less well known, holidays or interesting days that fall on Thursday approximately every seven years, and I often devoted a special column to them. Some of those special days are discussed below. The selection is arbitrary, but it gives a good look at what the Sky Spy columns covered over many years.

VALENTINE'S DAY

Valentine's Day is special for Arizona because it was admitted to the Union as the forty-eighth state on this day in 1912. Arizona was the last of the contiguous continental states admitted to the Union. If we could go back to the night of February 14, 1912, in Arizona the Moon would be a waning crescent and not present in the evening skies, which would be really dark without the Moon and the light pollution that plagues us today. At 8:00 p.m., Orion the Hunter sits on the meridian, and Sirius the brightest star sits high in the southeast.

EASTER; PASSOVER

Easter is a moveable feast that is not fixed to our civil Gregorian calendar. It follows the cycle of the Moon. The actual determination of Easter is somewhat complex. It is the first Sunday after the paschal full Moon, the first full Moon on or after March 21, the approximate time of the vernal equinox or the beginning of spring. Easter is closely related to the Jewish Passover. Passover begins on the fifteenth day of the Jewish month of Nisan, which is a spring month of thirty days that usually falls in March–April in our civil calendar. Thus, Easter and Passover are linked to each other in a complex fashion based on both the Moon and the vernal (spring) equinox. The cycles of the Moon and Sun were critical to most cultures until a few hundred years ago when accurate timepieces became available. The Sun and the Moon were important for religious reasons as well as for such practical purposes as determining when to mark the beginning and the end of the day, the month, the season, and the year.

IDES OF MARCH

Julius Caesar was told to "beware the Ides of March." This famous warning took place in Shakespeare's play *Julius Caesar*. Caesar was assassinated on the Ides of March, 44 BC. Interestingly, Tsar Nicholas II of Russia gave up his crown on this day in 1917. Hopefully, there will not be any assassinations on future Ides of March, or at any other time. In modern times the term "Ides of March" has become a metaphor for impending doom. However, the Ides was simply a term the Romans used for the fifteenth day of the months of March, May, July, and October. It was the thirteenth day of the eight other months. Nowadays, we are not superstitious and do not fear the Ides of March.

MEMORIAL DAY

Memorial Day originated as Decoration Day to honor Union and Confederate soldiers following the Civil War, when it became customary in May for family and friends to visit and decorate the graves of fallen soldiers. It was first observed on May 30, 1868, and I grew up with Memorial Day on May 30. It now honors all Americans who have died in all wars, and it officially occurs on the last Monday in May. Memorial Day is considered by many to be the start of summer, with Labor Day the end of summer. While Memorial Day is filled with parades, picnics, and other fun activities, we should all pause to remember those whose sacrifice helped preserve the blessings of our great nation. The stars and planets we see today also shone on those who served on foreign battlefields and the terrible battlefields of the Civil War, the costliest war in our history in terms of the dead and wounded.

If we could go to Gettysburg, Pennsylvania, on the evening of May 30, 1868, the first Memorial (Decoration) Day, at 9:00 p.m., we would see the gibbous 9½-day-old Moon high in the southwest. Scorpius the Scorpion is nearly completely risen in the southeast, Virgo the Virgin sits along the meridian high in the south, and above Virgo nearly overhead is Boötes the Herdsman.

D-DAY

D-Day is a generic term, though it is mostly understood to mean June 6, 1944, when the Allied invasion of Normandy took place in the Second World War. This was a combined naval, air, and land assault on Nazi-occupied France. World War II was a seminal experience for people of my parents' generation, and I grew up at a time when almost all adults had had definitive wartime experiences of some type.

Now, there are few veterans left from that time, and it is worthwhile to note that the weather postponed the planned June 5, 1944, D-Day invasion by twenty-four hours. If it had not taken place on June 6, the invasion would have been postponed at least two weeks to meet the requirements for decent weather, Moon phase, and tide phase. A full Moon was desirable because it would provide the best visibility for aircraft pilots and have the highest tides. The first assault landings were scheduled for just prior to dawn midway between low and high tide, with the tide coming in to improve visibility of obstacles on the beach and to minimize the time it took for the landing craft to make it to the beach to discharge the soldiers under extreme enemy fire.

Those soldiers in Normandy who made it off of the beach safely and had time to look at the night sky on D-Day at 9:00 p.m. that night would have seen the full Moon low in the southeast just to the left (north) of rising Scorpius the Scorpion, with bright red Antares at the heart of the Scorpion.

THE FOURTH OF JULY

July 4, Independence Day, is a national holiday in the United States. It is usually celebrated with fireworks in the early evening. If we could transport ourselves back to July 4, 1776, we would find darker skies and fewer people in Tucson, Arizona. Tucson had just been founded in 1775 and was part of Spanish Colonial Mexico. There was no light pollution, though summer days were very hot and dusty indeed without air conditioning. On July 4, 1776, the Moon was just past full, rising at 8:20 p.m. and setting the next day at 9:42 a.m.

We don't have daylight saving time in Arizona, for which I am personally happy. Nevertheless, one of our American heroes, Benjamin Franklin, is credited with first suggesting the idea in 1784. In 1776, the source of the Sun's heat and light was unknown, and no one knew the distances to other stars. The nature of stars themselves was unknown, though some guessed they may be faraway suns. What would Benjamin Franklin think if he were transported to our times? Would he be appalled by our wars and environmental abuses, or would he marvel at our scientific knowledge and technology? No one knows, but we do know he would recognize the same familiar friends in the night sky.

BASTILLE DAY

July 14 is Bastille Day, a French national holiday that officially commemorates the 1790 Fête de la Fédération, held on the first anniversary of the storming of the Bastille on July 14, 1789. The storming of the Bastille, a medieval fortress and prison in Paris, is traditionally considered the start of the French Revolution. The fortress was liberated from its guards in the late afternoon. One of the few advantages the average Parisian had in those days was dark skies. There were no electric lights, and the skies were dark even in the centers of major cities. Observatories were in large cities, there being no necessity to place them away from city lights. That changed by the end of the 1800s, when city lights, smoke, haze, and poor viewing conditions convinced professional astronomers to locate their modern instruments on mountaintops for the clearer, darker, and steadier skies required by large telescopes.

On the night of July 14, 1789, wherever it was clear in France, those who observed the sky would have marveled at the summer Milky Way high in the east and been delighted by a beautiful last quarter Moon rising shortly before midnight. Today, the citizens of France enjoy far better freedoms and economic conditions than their eighteenth-century ancestors, but they still enjoy the same stars and constellations.

APOLLO 11 MOON LANDING

On July 20, 1969, Apollo 11 landed on the Moon, and Neil Armstrong was the first person to walk on the Moon. Each generation has one or two days that everyone remembers. Sadly, most of these days commemorate terrible events, such as September 11, 2001, or President Kennedy's assassination. July 20, 1969, on the other hand, is a happy day. It represents the fulfillment of thousands of years of human wonder and endeavor.

On that fateful day, I was with a now-departed friend, Mike Nightingale, in Jasper National Park in Alberta, Canada. That morning we saw a chalkboard in the lobby at the resort in Lake Louise, Alberta, with the message, "The Eagle has landed." The Lunar Module had safely landed on the Moon's surface. Inside were Neil Armstrong and Buzz Aldrin. Michael Collins remained in lunar orbit in the Command Module.

Later on in the afternoon, we lunched in Jasper while watching a small black-and-white television set up in a large restaurant dining room. Probably a hundred people cheered when Neil Armstrong exited the Lunar Module and stepped onto the surface of the Moon. A Mr. Smith from St. Louis bought drinks for everyone in the house, and a good time was had by all. A lot has happened since then, some good, some bad, but the memory of that day is happy indeed for millions around the world.

THE DOG DAYS OF SUMMER

The hottest, muggiest days of the year are in July and August for most of the Northern Hemisphere. They are often referred to as the "dog days of summer." In common folklore, dog days means a period of stagnation or inactivity. When I think of dog days, I remember my old childhood dog Bozo lying around on a blistering Illinois August afternoon under the shade of one of our elm trees. However, dog days is a very ancient term with an astronomical origin. The Greeks and Romans called the hot days of summer dog days (*dies caniculares* in Latin).

Sirius the "Dog Star" is the brightest star and lies in Canis Major the Greater Dog. In ancient times, Sirius rose on hot summer mornings just before the Sun. Due to the slow spin of Earth's axis over a twenty-six-thousand-year period known as the precession of the equinoxes, Sirius now rises later in the summer than in ancient times. However, it still retains its link to hot summer days of August. In mid- to late August, you can catch Sirius rising in the dawn twilight prior to sunrise. Give it a try. Use a planisphere to estimate when Sirius rises, or, better yet, use an app on your computer or smart phone to determine the rise time of Sirius and the time of sunrise. Find a clear eastern horizon and look for Sirius to pop above the horizon halfway between directly east and directly southeast. If you are lucky enough to see Sirius early on a muggy morning, you can see why the Romans felt Sirius was the cause of hot, sultry weather. It's somewhat ironic that we now consider Sirius one of our winter stars, because we are used to best seeing it in all its blazing glory on cold winter evenings after sunset.

VETERANS DAY

In the United States, November 11 is celebrated as Veterans Day. In other parts of the world, it is known as Armistice Day or Remembrance Day. November 11 is the anniversary of the signing of the Armistice that ended World War I, with formal hostilities ending at the eleventh hour of the eleventh day of the eleventh month of 1918.

On the early morning of November 11, 1918, those soldiers and sailors up before sunrise would have been greeted by Venus hovering just barely above the eastern horizon. I am sure many of them hoped Venus would finally shine on a world with a peace that was due to start in a few short hours. Later that evening if the weather permitted, they would have seen Mercury just above the western horizon and the first quarter Moon high in the western sky. The Moon, for the first time in four years, was shining on a world at peace. Unfortunately, the First World War, the "war to end all wars," did not produce a lasting peace, but the stars and planets that greeted our grandparents and great-grandparents then are still following the same paths and seasons. We can't thank veterans for the stars and planets, but we can thank all veterans for many of our nation's blessings.

THANKSGIVING DAY

Thanksgiving Day is an important national holiday in the United States and Canada. A day of thanksgiving is also celebrated in several other countries, though it may be named differently. Thanksgiving in the United States is celebrated on the fourth Thursday in November. In Canada, Thanksgiving is celebrated on the second Monday of October.

Since my Sky Spy columns are published on Thursdays, every year I have a Thanksgiving column. I am always thankful for a wonderful family and friends. Every year I remind myself that, no matter one's political or religious beliefs, the vast majority of us feel there are countless things to be thankful for in this great nation. Of things to be thankful for close to home, I am very thankful that Tucson is the astronomy capital of the world and for our relatively dark city skies compared to most urban areas. I am thankful for the University of Arizona's world-leading astronomy, planetary science, and optical sciences programs, for the superb professional observatories in Arizona, and for the wonderful Tucson Amateur Astronomy Association, Inc. (TAAA), our local amateur astronomy club. The Planetary Science Institute (PSI) and the International Dark-Sky Association (IDA) add much to Tucson's astronomical heritage.

If we could go back to 1621, the year of the Pilgrims' first Thanksgiving celebration, and assume it was on November 25, we would notice the Moon is up most of the night, rising at 3:06 p.m. in the afternoon and not setting until 3:28 a.m. the next morning. Venus would be low in the southwest after sunset, while Jupiter would be rising in the east, becoming more prominent as the night went on. The land and people have certainly changed since those days, but the stars and planets would have been as familiar to us then as they are today.

ASTRONOMY DAY

International Astronomy Day is a semiannual event in which astronomy enthusiasts, professional and amateur, interact with the public. The theme of Astronomy Day is "bringing astronomy to the people." Astronomy Day is a worldwide event celebrated each spring and fall. A general overview of the event can be found on the Astronomical League website at: https://www.astroleague.org/al/astroday/astrodayform.html.

GLOBE AT NIGHT

Globe at Night is an international citizen-science campaign to raise public awareness of light pollution. Everyday citizens are asked to measure the night-sky brightness and submit their observations to the Globe at Night. There are selected campaigns every year to look at a particular constellation to see how many stars one sees, which gives an indication of how dark the observer's sky is at that location and time. Go to http://www.globeatnight.org/ for more information.

8

Selected Famous Astronomers, Events, and Places

INTRODUCTION

O CCASIONALLY, THE column would run on an important astronomical anniversary date, such as the launch of Sputnik 1 or the birthday of William Herschel. I often devoted at least a part of the column for that day to the person or event whose anniversary date is coincident with the column. This led to a good collection of interesting columns about astronomers, astronomical events, or astronomical places. The selection of the people, events, or places was very arbitrary, mainly dependent on the timing of the Sky Spy columns. While the selection process was random, it did produce an overview of astronomy.

AMERICAN ASTRONOMICAL SOCIETY

The American Astronomical Society (AAS) was established in 1899. The AAS membership of more than eight thousand scientists includes astronomers, physicists, mathematicians, geologists, engineers, and others whose work and research lie within a wide range of subjects comprising modern astronomy. It is the organization for professional astronomers in the United States. According to its website, "the mission of the American Astronomical Society is to enhance and share humanity's scientific understanding of the universe as a diverse and inclusive astronomical community." I consider the AAS to be one of the premier scientific organizations in the world. Check out its considerable activities at: http://aas.org/.

YURI GAGARIN

Yuri Gagarin (1934–1968), the first human in space and the first to orbit Earth, was launched into space on April 12, 1961, aboard Vostok 1. He spent one hour and forty-eight minutes in orbit and is reported to have said from space, "The Earth is blue . . . How wonderful. It is amazing." Tragically, Gagarin and his flight instructor Vladimir Seryogin died on a routine training flight in a MiG-15UTI on March 27, 1968. Gagarin was highly respected worldwide for his brave accomplishment and for his simple, unpretentious demeanor.

GENERAL RELATIVITY

This book is not the place to discuss relativity, except to note an important anniversary associated with the scientific community's acceptance of the theory. On May 29, 1919, there was a total eclipse of the Sun. Precise measurements taken during that eclipse by Sir Arthur Eddington (1882–1944) helped to confirm a key prediction of Einstein's theory of general relativity that the path of light passing close to a massive body will be bent slightly.

Eddington observed the eclipse from the island of Príncipe near Africa. He carefully measured the positions of stars in the Hyades cluster in Taurus visible near the Sun's edge during the eclipse and determined that their apparent positions had been shifted by nearly the amount predicted by relativity. These stars could only be observed during the eclipse, because ordinarily they would be hidden by the Sun's glare.

When Eddington announced his results the following year, Einstein became an overnight sensation and acquired his legendary stature that endures to this day. Eddington helped cement his own reputation as one of the greatest astrophysicists of all time, though some of his best and some of his more controversial work was yet to come. It is sometimes said that, from approximately 1914 to 1924, in the entire world only Arthur Eddington and Albert Einstein understood relativity. Eddington was also the first scientist to develop a true explanation for the inner workings of stars.

DORRIT HOFFLEIT

Ellen Dorrit Hoffleit, a world-renowned research astronomer at Yale University, was born on March 12, 1907, and died April 9, 2007, almost a month after turning one hundred. She is widely recognized for her editorship of the *Bright Star Catalogue*, an important listing of the nine thousand brightest stars.

More importantly, Hoffleit is recognized for her mentorship of generations of astronomers, including many women. She earned her PhD in astronomy from Radcliffe College in 1938 and later spent most of her career at Yale University before retiring in 1975. From 1957 to 1978, she served as the director of the Maria Mitchell Observatory on Nantucket and developed the summer research program for students there, which still runs to this day. She was an incredibly passionate teacher.

The AAS awarded her with the George Van Biesbroeck Prize in 1988 in recognition of her lifetime of service to astronomy. I was most fortunate and honored to meet her when she came to Tucson to receive the prize.

INTERNATIONAL ASTRONOMICAL UNION

The IAU was founded in 1919 and, according to its website, "its mission is to promote and safeguard the science of astronomy in all its aspects, including research, communication, education, and development, through international cooperation." The IAU is the international organization devoted to professional astronomy and is recognized by astronomers worldwide. It sets many of the standards used by professional astronomers, including the formal establishment of the constellation boundaries and names, as well as the names for surface features on the Moon, planets, and other astronomical bodies.

INTERNATIONAL DARK-SKY ASSOCIATION

The IDA was founded by Dr. David Crawford, a professional astronomer, and me, an amateur astronomer, in 1987. It was incorporated that year as a tax-exempt nonprofit corporation whose mission is to combat the growing threat of light pollution and to advocate for quality nighttime lighting to preserve dark skies. The IDA is based in Tucson, Arizona, and has a worldwide reach. It fully supports proper nighttime lighting for public safety, security, and recreation. Such lighting should be used only when necessary and should be designed to minimize light pollution and **light trespass**. More information about the IDA can be found at: https://www.darksky.org/.

MARCH 13; LOWELL OBSERVATORY

March 13, 2009, was a Friday the thirteenth, which led to a discussion of Friday the thirteenth and to many important anniversaries that happen to fall on March 13. Friday the thirteenth in and of itself has no astronomical significance. Ancient superstitions make us all a little more cautious whenever a Friday the thirteenth occurs, and that has led to a series of blood-curdling movies.

However, March 13 is an anniversary date for several important astronomical events. William Herschel discovered Uranus on March 13, 1781, a Tuesday. His discovery of a new planet expanded the number of known planets beyond the six **classical planets** (Mercury, Venus, Earth, Mars, Jupiter, and Saturn). Uranus was also the first planet discovered using a telescope.

Percival Lowell (1855–1916) was born on March 13, also a Tuesday. Lowell founded Lowell Observatory in Flagstaff, Arizona, in 1894. Lowell Observatory is a wonderful astronomical institution for which all Arizonans take pride. It is a private, nonprofit research center with a long colorful history of astronomical discovery and research. Lowell Observatory is best known to the public as the place from which Clyde Tombaugh discovered Pluto in 1930. However, it remains on the cutting edge of discovery and research in many areas. Check out its web page at http://www.lowell.edu/. Also, be sure you take time to tour the observatory whenever you are in Flagstaff.

On March 13/14, 1986 (Thursday/Friday), the European Space Agency's Giotto probe took close-up pictures of Comet Halley's nucleus. Giotto was named after the medieval Italian painter Giotto di Bondone (ca. 1267–1337), who observed Comet Halley in 1301. Information from Giotto and the Soviet probes Vega 1 and Vega 2 confirmed comets are large "snowballs" composed of a complex mixture of frozen water, carbon monoxide, methane, ammonia, and other compounds.

CHARLES MESSIER

June 26 is the birthday of Charles Messier, a famous eighteenth-century astronomer. He bridged the transition from purely observational astronomy with small telescopes and little instruments to astronomical research involving larger telescopes, better optics, and the introduction of instruments for measuring star positions and star spectra.

Messier's career spanned the time of the French monarchy through the French Revolution and the Napoleonic Wars. He was highly respected in European scientific circles for his observational skills and for his famous catalog of important sky objects.

Messier was mainly interested in discovering comets and made his now-famous list of objects so others would not confuse them with comets. Messier's list contains some of the brightest star clusters, galaxies, and nebulae in the sky. Amateur astronomers spend endless hours viewing and photographing the **Messier objects**. A few of Messier's objects are visible to the naked eye, and many of them can be seen with binoculars at a dark-sky site.

MARIA MITCHELL

Maria Mitchell (1818–1889), a leading nineteenth-century astronomer, was born into a prominent Quaker family on Nantucket Island off the coast of Massachusetts. She had nine brothers and sisters, who were raised by their parents with access to quality education for all. Maria showed a talent for mathematics and astronomy and learned to use her father's telescope at a young age. On October 1, 1847, she discovered a comet using a small three-inch telescope and cemented her discovery claim by calculating its orbit, a considerable mathematical feat.

Maria Mitchell's lifetime of observing achievements enabled her to become professor of astronomy at Vassar College in 1865 and director of the Vassar College Observatory. At that time, Vassar was a women's college, which became coeducational in 1969. Maria Mitchell was world famous during her lifetime and received many honors because of her and her students' years of observations and discoveries. However, she is mostly noted for her lifelong struggle for women's rights and for being a strong opponent of slavery.

NATIONAL AERONAUTICS AND SPACE ADMINISTRATION

The National Aeronautics and Space Administration (NASA) was established on July 29, 1958, when President Eisenhower signed the

National Aeronautics and Space Act, replacing NASA's predecessor, the National Advisory Committee for Aeronautics (NACA). NASA is one of the most recognized acronyms in the world, and it is one of the most accomplished and recognized governmental agencies in the world. Technically, it is an agency of the executive branch of the United States government.

NASA is a big organization that spends vast sums of money and has its ups and downs, but its list of accomplishments is amazing. A few of NASA's more stellar accomplishments are the Apollo missions to the Moon, the Skylab space station, the Space Shuttle, the International Space Station (along with our international partners), and its many scientific missions to the planets, including Mercury, Venus, Mars, Jupiter, Saturn, Ceres (a dwarf planet), Vesta (an asteroid), and Pluto (a dwarf planet).

CECILIA PAYNE-GAPOSCHKIN

Cecilia Payne (1900–1979) was born in Wendover, England. Despite being extraordinarily bright and well educated in physics and mathematics, she had little opportunity for an astronomy career in the United Kingdom because she was a woman. She moved to the United States in 1923 at the recommendation of Harlow Shapley (1885–1972), director of the Harvard College Observatory.

In 1925, she became the first person to earn a PhD in astronomy from Radcliffe College (now part of Harvard). Her PhD thesis has been called the most brilliant ever written in astronomy. In it she established that hydrogen is by far the most abundant element in stars and the most abundant element in the universe, an amazing conclusion that went against conventional thinking at the time.

Cecilia Payne married the Russian astrophysicist Sergei Gaposchkin and they had three children. She had a long, distinguished career at Harvard University, and she was the first woman to head a department there. Whenever you enjoy the starry sky, think of Cecilia and the fact that most of the universe is hydrogen, with some helium thrown in and with mere traces of the everyday elements we find on Earth, all of which she discerned when she was a graduate student.

PLANETARY SCIENCE INSTITUTE (PSI)

The PSI, based in Tucson, Arizona, is dedicated to the exploration of the solar system. It was established in 1972, and according to its website it is the largest private (nongovernmental) employer of planetary scientists worldwide. The planetary scientists at PSI are world leaders in planetary exploration, and I had the wonderful experience of being on its Board of Trustees for many years. For more information about PSI, see: https://www.psi.edu/.

SPUTNIK

Sputnik 1, the first artificial satellite to orbit Earth, was launched October 4, 1957. Unlike the happy memory of the Apollo 11 landing on the Moon, the memory of Sputnik's launch is bittersweet for me. I was in junior high school at the time, and I can still see the headlines of the *Chicago Tribune* proclaiming the Soviet Union had just put the world's first artificial satellite into orbit. The memory is bitter because I was disappointed the Soviet Union beat us to the punch. The memory is sweet because I was vaguely aware that Sputnik signaled the start of the Space Age, something I was fortunate enough to enjoy and witness firsthand from its earliest stages.

Little did I realize then that the Soviet Union would no longer be around in my middle age and that space exploration would someday be taken for granted. Every clear night, it is now quite easy to see one or more satellites as they pass overhead in the evening or morning twilight. I also didn't realize then that the dark skies I took for granted growing up in the suburbs of Chicago would become an endangered species from light pollution.

Sputnik, which means "satellite" in Russian, was the size of a basketball and weighed only 183 pounds. It orbited Earth every ninety-six minutes and completed almost 1,500 orbits before it burned up in Earth's atmosphere on January 4, 1958. Compared to today's satellites, Sputnik was very primitive, but it was revolutionary in its time. Whenever I see a satellite in the night sky, I try to reflect on those mostly unsung engineers, scientists, and ordinary workers who put Sputnik into the sky. No doubt most of them are now gone, but mankind surely owes them a tremendous debt of gratitude.

CLYDE TOMBAUGH

Clyde Tombaugh, discoverer of Pluto, was born on February 4, 1906. Even though Clyde was born in Illinois, went to college at the University of Kansas, and spent much of his later professional career at New Mexico State University, Arizona can also claim him as one of its own. He discovered Pluto at Lowell Observatory in Flagstaff and spent many years in Flagstaff. I was most fortunate to know Clyde and his beautiful wife, Patsy. Although he was a world-famous scientist, he was a delightful person with a sparkle in his eye and a love for crow jokes.

9

What I Learned from Writing an Astronomy Column

INTRODUCTION

WHEN YOU write a newspaper column about "what you can see in the sky tonight," you are presumed to be an expert and to know more about the subject than your audience. If you take such an endeavor seriously, you may or may not produce a good article, but you will certainly learn a lot along the way. It is commonly accepted that having to write an article or give a talk about a topic is a great learning experience. Having to explain a concept to someone else means you must understand it as well. I have often been chagrined to discover how little I knew about an astronomical topic when I first sat down to write a column about it. When I finally finished the column, it might not have been any good, but I sure learned a lot.

Putting together in simple language an explanation for a phenomenon such as the phases of the Moon is very hard work and a brutal learning experience. If nothing else, you learn quite a bit, and if all goes well, so does your audience. The Sky Spy columns continually provide such an experience for me. Hopefully, the readers enjoy the columns and learn about the sky. No matter what, I am challenged and constantly learning.

This chapter reflects upon my experience writing Sky Spy, mainly emphasizing the errors and omissions in, as well as reader feedback from, the columns. In this way, I hope you appreciate what one should and should not do when undertaking a similar endeavor.

COMMON ERRORS

I carefully proofread every Sky Spy column, and I have carefully proofread this book. I have taken great pains to make sure the book is not a clumsy compilation of old newspaper columns. This required quite a bit of work sorting through the columns and correcting errors I made along the way. Some of the errors were substantive, with erroneous or confusing information presented to the reader. Many of the errors were simple typos missed by my proofreading. Proofreading is no fun. It takes as much time for me to proofread an article as it does to initially write it. Every Sky Spy column is thoroughly vetted by the editor of the *Caliente* section of the *Arizona Daily Star* newspaper. The editor composes the headline for each column using material from the column to make a catchy phrase that is supposed to attract the reader's attention.

I have no say in these headlines, and they are usually pretty good, much better than I would write. However, once this caused an embarrassing error. The column for August 21, 2008, was about the summer triangle of Altair, Deneb, and Vega. In the column I noted this "triangle" is an asterism of three bright stars. The headline for the column that day was, "Summer Triangle is an asterisk in the sky" (figure 9.1). The editor had confused the word asterism with asterisk and composed a headline accordingly. When I pointed this out to her, she was embarrassed about it, as was I, but, frankly, only a very astronomically learned reader would have caught this error, and I never heard about it from any reader. In fact, the headline was rather catchy.

www.aznightbuzz.com

SKY SPY

Summer Triangle is an asterisk in the sky

By Tim Hunter
SPECIAL TO THE ARIZONA DAILY STAR

Now is a good time to look for the Summer Triangle, which is an asterisk of three bright stars, Altair, Deneb and Vega.

An asterisk is merely a grouping of stars that's not an official constellation.

Altair is in the constellation of Aquila the Eagle, Deneb is in Cygnus the Swan, and Vega is in Lyra the Lyre. In the early evening they are directly overhead.

While you're at it, look for the minor constellations Vulpecula the Fox, Sagitta the Arrow, and Delphinus the Dolphin.

It will take a dark sky to see these smaller constellations well, but the stars of the Summer Triangle are readily visible if you know where to look.

The Summer Triangle is reasonably easy to recognize because its stars are among the brightest in the sky.

Altair is 17 light years away and is a very strange star that's rapidly rotating, having an ellipsoid shape like an egg.

Deneb is so far away its distance is not known with any certainty, but it's estimated to be roughly 3,000 light years from us. Even so, Deneb is the 19th-brightest star in our sky and probably one of the most luminous stars in the entire Milky Way.

Vega is the fifth-brightest star in the sky and is about 25 light years from us. Vega is surrounded by a large disk of gas and dust that possibly contains planets.

The moon: It's in a waning (getting smaller) gibbous (more than half lit) phase. It will be at last quarter on Saturday and new moon on Saturday.

Tim Hunter has been an amateur astronomer since grade school. Contact him at skyspy@azstarnet.com.

FIGURE 9.1 Sky Spy column for August 21, 2008. The column headline should have said "Summer Triangle is an asterism in the sky."

A common and troublesome error is to omit an important word in your writing but subconsciously put it in your thoughts when you are proofreading. At best, most such mistakes are caught, and the errors are minor, leaving out an article or preposition that the editor or reader can mentally insert for themselves. Sometimes a very important word is left out, and it is not apparent to the reader or editor.

Everyday email and letters are filled with typographical mistakes, obscure abbreviations, dropped words, incorrect dates, incorrect places, and many other miscellaneous mistakes. Usually, these are unimportant and overlooked by the reader. Abbreviations, acronyms (short words or names formed from a longer official name), and eponyms (a discovery or phenomenon named after the person who presumably discovered or first described it) are common to medicine, law, and science. NASA is the abbreviation or acronym for the National Aeronautics and Space Administration. The Messier numbers are an eponymous way to identify those celestial objects discovered or described by the famous astronomer Charles Messier in the eighteenth century.

Abbreviations are common in professional and amateur astronomy and can be most difficult and annoying for someone not involved in the discipline day to day. In this book I have taken extreme care to use as few of these as possible and to define them precisely. The Glossary lists and defines terms used frequently in everyday astronomy. It is always better to err on the side of using the proper name for something rather than the abbreviation, writing "International Astronomical Union" instead of "IAU," for example.

Certainly, no abbreviation out of the ordinary should be used without first being defined. It is reasonable to assume the reader knows such common abbreviations as USA (United States of America), EDT (eastern daylight time), and NASA, especially from the context in which they appear. It is unreasonable to assume the average reader understands that HST means the Hubble Space Telescope, or that ISS means the International Space Station. Most readers may recognize terms like the ecliptic or the zodiac without really understanding them. I take care to define such words every time I use them.

I learned not to bother publishing corrections (errata) for simple column errors or omissions. Important errors such as incorrect dates or times for astronomical phenomena certainly must be corrected as soon as possible. Today, the *Arizona Daily Star* has an online edition as well as the usual printed edition. I am not sure of the overlap in readership between the two. Theoretically, there could be daily online Sky Spy columns, but the weekly printed/online version has worked well enough.

EMAIL–QUESTIONS, COMPLAINTS, AND CORRECTIONS

My personal email is not listed with the column. The reader is directed to address any email to me through the editor of *Caliente*, who forwards it to me. This works well, and I receive one or two messages a month concerning the column. Usually, they are questions or compliments. A few are complaints or important corrections that point out serious errors on my part.

COMMON EMAIL QUESTIONS AND THEMES

The most common email questions have been:

1. What is that bright star?
2. What was that bright light I saw last night?
3. You said the Moon was at first quarter, but it was half lit.
4. Why do stars twinkle and planets don't?
5. When can I see the full Moon?
6. What telescope should I buy?
7. When can I see the International Space Station?
8. Why isn't your column listed in the index for *Caliente* or in a consistent place?
9. Is Mars soon going to look as big as the full Moon?
10. Where can I see the Milky Way?
11. Who should I give my never-used telescope to?
12. We are planning a trip to Arizona. Where should we go and what should we bring with us?

Most of the time, the bright star a reader has observed turns out to be Venus. Sometimes it is Jupiter or Sirius. If the reader's question is detailed enough, I can almost always identify the bright star in question. That said, I cannot usually identify a brief bright flash of light seen by a reader, particularly if the time and location of the light flash are poorly described. I can offer a few suggestions depending on how well the reader describes their experience. A bright meteor or fireball is likely in some of the situations. Sometimes, the ISS is very bright and surprises the inexperienced observer.

I mention the first and last quarter Moon quite often. Most readers seem to realize a quarter Moon is half lit from our perspective, but I do have to take care to explain that "quarter" is a term used in reference to the Moon being one quarter or three quarters of the way through its orbit from a preceding New Moon. It does not mean we see the Moon quarter lit.

Most bright stars twinkle, particularly near the horizon, and usually the planets show no obvious twinkling. I never found this very satisfying when trying to identify whether a bright "star" is truly a star or is a planet. I feel it is best to just learn the planets and brighter stars and not worry about twinkling versus not twinkling.

I have received surprisingly few questions on what telescope to buy, and I have not really covered the topic in the Sky Spy columns. The next chapter explains my recommendations regarding telescope purchases. In a couple of instances, readers have offered to donate seldom-used telescopes to a deserving student or astronomy club. I have put the reader in contact with the appropriate person at the TAAA who can best decide if the telescope is worthy of use and who would most benefit from its receipt.

A weekly newspaper column is not a good place to discuss satellite observations and predictions (see below). Every August it seems there is a widespread web rumor that Mars is approaching Earth and will be as large as the full Moon. That is nonsense and not true, and when I receive email from friends passing along this myth, I explain this is a common urban legend that has had quite a long life.

Inquiries or complaints about how the column is listed or displayed in the *Caliente* section of the *Arizona Daily Star* have been infrequent, though I use the occasion to refer the reader to the editor of *Caliente* in the hope that such a complaint will get me a more consistent placement in the newspaper or a bigger column spread. So far, that has not worked.

The Milky Way is discussed throughout the year in the Sky Spy columns, with directions on where to look for it in the sky and with a recommendation to get out of the city away from urban light pollution. Since I live in Tucson, Arizona, and the Sky Spy column is in the *Arizona Daily Star*, I sometimes get questions on what to see and do, astronomically speaking, for someone visiting Arizona for the first time. The answer varies depending on the reader's specific question and experience regarding astronomical observing. I have put together a resource list for astronomy-related questions and topics (see the Astronomical Resources section at the end of this book), some of which pertain particularly to Arizona. I often forward this to an inquiring reader.

EMAIL COMPLAINTS AND CORRECTIONS

Almost all of the complaints I have received about the Sky Spy columns have been valid. In fact, I wonder why there are not more. I guess those who feel the column is a waste of time don't bother to write me. They simply stop reading the column.

One of the earliest and probably most important complaints I received was that I did not adequately identify which direction to look for an object discussed in a column. Since diagrams were included with the columns until 2009, I was too casual about giving sky directions. Now, I am most careful to constantly make sure I adequately inform the reader as to when and where to look for an object, being careful to note the direction of the object (east, west, north, or south) and how far it is from the horizon.

I also learned to describe astronomical phenomena in simple, everyday terms, and when using astronomical words, to define them in simple language. In many cases, I define astronomical terms like gibbous, waxing, waning, and elongation every time I use them. One cannot assume that a given reader has read any previous column or remembers a past definition from a column weeks or months earlier.

One reader was very upset by my describing an object as being so many degrees above the horizon. This reader did not understand that the sky stretches 180 degrees from one horizon to another, that an object overhead is 90 degrees from the horizon in any direction, and that an object halfway up from a specific horizon has an altitude of 45 degrees. Most readers probably understand these concepts, but I now take care to define "degrees of altitude" every so often, so a reader who is new to the column or to observing the night sky understands when I say "look directly west 45 degrees above the horizon." I also mention that the Moon has a diameter of 0.5 degree and note common ways to use one's hand and fingers to guess separations of 1, 5, and 10 degrees.

Some readers know me personally and send me direct email comments. I am meticulous in responding to all readers' questions or complaints. Now I often send a reply directly back to the reader and have not had any problem with this, even though it gives a stranger my personal email address. I used to rely entirely on the editor to forward my answers back to the reader, but I learned these sometimes got lost, and the reader may have never received a reply.

Many years ago, a schoolteacher wrote me inquiring about how to look up rise and set times for the Moon and its phases. She wanted this information to help teach her middle school science class. I wrote her a lengthy reply and even offered to buy her a copy of the wonderful program LunarPhase Pro. I sent this message back through the newspaper and never heard from the teacher. When I later inquired about this message and my reply, the editor remembered it, but all of the correspondence got lost in the newspaper's cyberspace, and we could never reconnect with the teacher. This led me to directly answer any emails I thought were especially important, sending a copy of my reply to the editor.

SATELLITE ERRORS

Satellites are ubiquitous. If the sky is clear, they are visible every night in the early evening or in the morning sky somewhat prior to sunrise. They are not visible in the middle of the night because they are too deep in Earth's shadow to receive any sunlight. Some very high altitude satellites may remain in sunlight later in the evening, but they are not bright enough to be visible to the naked eye. Sometimes they are evident going through the field of view of a telescope or binoculars.

In the spring of 2015, the editor of *Caliente* asked me to include a small separate section of the Sky Spy column for bright satellite observations. Several years earlier, I had been asked to make a separate section of the column for observing the Moon, its phase, rise or set time, or whatever was pertinent for the Moon for that week. This Moon section worked well and is a part of the Sky Spy columns to this day.

The section on satellite observations was a mistake, and I should have known better. Since I usually wrote Sky Spy columns three to four weeks ahead of their publication, satellite predictions that far ahead were often quite inaccurate. Even the ISS periodically boosts or modifies its orbit. A weekly newspaper column is not appropriate for satellite observations. I have always used Heavens Above (https://www.heavens-above .com/main.aspx) for my satellite observations. This website is the standard for predicting a satellite pass, its time, brightness, and sky path. I have used it for many years to identify a satellite I observed either shortly after the observation or the next day, depending on the circumstance.

For the column, I listed predicted times and paths for the ISS and HST. They are bright and frequently visible. Everyone enjoys the ISS when it is bright and goes across a large part of the sky. On a good pass, it is easily visible in a light-polluted sky. I took care to warn the readers that the predicted satellite observations would vary a bit due to their being in a somewhat different location than my home, which was the basis for the satellite predictions. I assumed most of the readers were in the Tucson metropolitan area.

The satellite predictions soon caused complaints from readers unable to observe what I had predicted in the column. Changes to satellite orbits, though small from day to day, are large enough to prevent a good prediction for low-lying satellites like the ISS or HST a few weeks ahead of time. The predictions proved to be several minutes off, large enough to be useless. Finally, I convinced the editor to omit this part of the column after an error-plagued six-month run.

I do occasionally talk about satellites in the Sky Spy columns. In this case, I mention Heavens Above and recommend that the reader refer to the website for satellite observation information and to not rely on predictions more than several days in advance.

LONGEST DAYS AND NIGHTS

There are several predictable columns every year: the spring and fall equinoxes, the summer and winter solstices, and the brighter meteor showers, the Perseids, the Leonids, and the Geminids. There is always a Thanksgiving column, and often a column or two about our calendar and a discussion of solar and lunar calendars. For several years, at the spring and fall equinoxes, I used to say that the hours of day and night were equal. In June 2012, a most perceptive reader named Bob L. wrote, "I think you may have written something this week that is a common misconception: the vernal and autumnal equinoxes do not have days and nights of equal length. I believe it's usually five to ten days from the equinoxes that the days and nights are usually equal. The Sun rises and sets due East and West on the equinoxes."

Bob was correct. As I discussed in chapter 7, the hours of day and night are not quite equal at the equinoxes. For Tucson's latitude, equal hours of day and night occur five to six days before or after.

The longest day of the year (time from sunset to sunrise) is at the summer solstice, and the longest night of the year (time from sunset to sunrise) is at the winter solstice. However, the lengths of daylight a few days before and after the solstices are approximately the same as on the solstice itself.

ASTRONOMERS

One of the most delightful emails I received was from Elana, an eighth grader who was doing a report on astronomers for her English class. To receive full credit, she needed to interview an astronomer. She considered me to be one and interviewed me by email, asking a variety of questions. This was a tough assignment. Here are the answers I gave to Elana (I hope they helped her report).

1. *What are the benefits of being an astronomer for a career?*

First, we must define what we mean by the term "astronomer." There are two types of astronomers—professionals and amateurs. An amateur astronomer enjoys looking at the sky, may have a telescope, and may take pictures of the sky or even do various scientific measurements that he or she will send to professional astronomers. An amateur astronomer makes no money from astronomy. A professional astronomer is someone with an advanced college degree in astronomy and gets paid for doing astronomical observations or research.

Astronomers study the "big picture," the universe. They enjoy the beauty of the night sky and the Sun and planets while studying them. They study things that are the largest, coldest, hottest, and most violent in existence. They also study the smallest things possible: atoms. Astronomers use mathematics, physics, and chemistry as tools, along with telescopes and computers.

2. *What is the average salary for someone working in your field?*

Their salaries can vary tremendously. An astronomer may work only part time teaching courses at a junior college or community college and earn $10,000 a year. Or an astronomer may run a large research institute or large observatory and have many grants. Such a person would earn over $200,000 per year. On average, astronomers are not that well paid and make less money than a lawyer or doctor, but they generally earn more than a schoolteacher.

3. *Describe a typical day at work.*

An astronomer's workday is variable. Some only teach, while others both teach courses and carry out scientific research. In other words, much of the time professional astronomers work during the day teaching courses, writing papers, and looking at research data. They apply for grants and time on professional telescopes. When they are awarded telescope time, they may travel to the observatory and stay up all night making observations at the telescope. They operate the instruments on the telescope to obtain the data they need (they don't look through the telescope), while a telescope operator working for the observatory controls the telescope.

4. *What would be a disadvantage of your career?*

5. *Are there various levels of the job? If so, what are they?*

6. *What classes should one take to be prepared for this job?*

The disadvantage of being a professional astronomer is the generally low pay for long hours of work and the many years required for education. A typical professional astronomer can have four years of college for a bachelor's degree, five years of graduate school for a PhD, and at least two years of postdoctoral work at low pay before their first independent professional job. There are also not many jobs for PhD astronomers, and many of them work in other industries because they are well trained in mathematics, physics, and computational science. Mathematics is a tool necessary for most astronomical work, and to be a good professional astronomer, one must be well versed in mathematics.

7. *Are many in your field female?*

Women currently make up about 30 percent of the astronomy workforce. The number of women receiving PhDs in astronomy is about equal to the number of men, but men still hold the majority of senior positions in the field. Professional astronomical organizations such as the AAS are working very hard to increase the number of women and minorities in the ranks of professional astronomers. There are many scholarships and other programs to encourage women and minorities to become scientists and engineers. This is a great time to be thinking about a career in science and engineering.

8. *What characteristics do you need to do well in this career?*

To do well in any career, you should have an inquisitive mind and be willing to work hard. To paraphrase Thomas Edison: genius is 10 percent inspiration and 90 percent perspiration. Willingness to spend hours studying and going beyond what is minimally necessary will help one be successful no matter what. Also, be respectful of others and be generous with your time and teaching. Have a good grounding in mathematics.

9. *What is the highest position in your career?*

Personally, I am an amateur astronomer. In the professional astronomer world, a high position would be a director of a major observatory and/or the head of an astronomy department at a leading research university. A high position would also be head of a major division of NASA or director of a free-standing large scientific research corporation. Some professional astronomers have become presidents of large universities, though in that case they are usually doing mostly administrative work and little professional astronomy.

10. *What is the lowest position that still deals with astronomy?*

The "lowest" position that deals with astronomy is that of an amateur astronomer who enjoys looking at the sky. He or she may own a telescope and may join an astronomy club. He or she often participates in star parties and shows the sky to other amateur astronomers or to the public. Everyday citizens who help support astronomy and related sciences in the schools and universities and through state and federal governments by paying taxes that support scientific grants, NASA, and the National Science Foundation can be thought of as amateur astronomers.

CONCLUSIONS

Much of the best material for the Sky Spy columns has come from its readers through questions or complaints. Constant reader feedback is essential for keeping a column fresh and relevant. Traditional newspaper columns are fading from public view due to the challenges facing print media in today's digital world. Even so, a blog or digital column needs as much input as possible from interested people, both readers and editors. The sky is a wonderful draw. It interests everyone in some fashion. Put a telescope on a busy corner in the heart of a metropolis. You will draw a crowd regardless of light pollution or the surrounding urban chaos. It is hard to beat the glory of Saturn's rings or the craters on the Moon viewed through a small telescope. The summer Milky Way overhead on a clear night at a dark-sky site rivals any digital trick available to the modern movie industry. A total eclipse of the Sun is such a stunning experience that it has no serious rival in nature.

10

What Telescope Should You Buy?

"**W**HAT TELESCOPE** should I buy?" None. Buy nothing until you have looked through many telescopes and compared what you can see with what you can afford. It is better to buy a planisphere or a mobile app. Learn the constellations and the planets. Do not rush out and spend a lot of money on a telescope. Many cheap telescopes have poor optics and are unusable. More expensive telescopes may have good optics but can still disappoint you. Your expectations may exceed what the telescope can deliver.

Even a large amateur telescope with a very sturdy mount and excellent optics will not give a view like that of Hubble Space Telescope images. They will give hours of enjoyment for those knowledgeable of their capabilities, and their enjoyment increases with increased use as you develop your observing skills. Buying an expensive telescope without having experience using telescopes or a knowledge of the sky is an automatic setup for disappointment. The telescope soon becomes an expensive white elephant sitting unused in the back of a closet. Nothing is so forlorn as an unused telescope hidden away and collecting dust.

Before considering buying a telescope, I recommend checking out a local astronomy club, which is present in most large cities. They are fun and worth joining. I have been a member of the TAAA for more than forty-five years. Astronomy clubs have periodic star parties, where members and guests gather after hours in a park or dark-sky location to view the night sky with like-minded fellows. Star parties are almost always free and open to the public. What better way is there to introduce the hobby and the local astronomy club than with a star party under the starry sky with telescopes for observing? Awe and wonder at a star-filled sky is part of our human makeup.

At most amateur astronomy star parties, there are several telescopes of various sizes and designs pointed at the Moon, stars, planets, star clusters, nebulae, or galaxies, and you are free to wander around and look through them. This is a wonderful way to learn about the night sky and to learn about telescopes. I highly recommend attending as many star parties as you can.

An experienced amateur astronomer knows what telescope they want to buy. It may or may not be in their price range, but they have enough experience to know what to expect and what to do with it. It is a big mistake for a beginner to buy a telescope without taking the time to see what is available and think through exactly what they want to do with it. Instead, an excellent investment is a pair of binoculars with good optics. Buy binoculars that you can take to a football game and enjoy using. They should not be too big or heavy and should be easy to hold for a long time. If they work well at a football game, they will work well on the night sky and give you hours of pleasure. Again, members of a local astronomy club can help you pick out a well-rounded pair of binoculars for both daytime and nighttime use.

The best way to learn a new hobby is to have an experienced buddy willing to spend lots of time with you as you get started in the new adventure. Amateur astronomy is no different. Having a good amateur astronomy buddy is a blessing. One of the best ways of finding amateur astronomy buddies is to join the local astronomy club, attend their meetings, and go to as many star parties as possible. Do not be shy. Most astronomy club members are beginners themselves or remember when they first began the hobby. Most hobbyists are friendly and want everyone else to enjoy their passion as much as they do.

If you live in an area devoid of an astronomy club, there still may be one or more local amateurs that can be sought out. The internet is mostly a blessing in this case, offering a worldwide choice of venues at which to learn about astronomy, telescopes, and viewing the night sky. Read about the Astronomical League, the largest confederation of astronomy clubs in the United States. If you live in Canada, check out the Royal Astronomical Society of Canada (RASC). If you are more internationally minded, look at Astronomers Without Borders. The Astronomical Resources section at the end of this book lists many organizations and tools for you to explore.

Consider either a printed or digital subscription to leading amateur astronomy publications like *Astronomy*, *Sky and Telescope*, *Amateur Astronomy*, and *Astronomy Technology Today*. I am hesitant to list many magazines, astronomy shops, or websites, as no doubt I will overlook something important, and I realize a book such as this cannot always provide timely information. Do a bit of homework and spend time on the internet to explore astronomical resources. It will be fun and quite rewarding.

In earlier times, experienced amateur astronomers decried the "department store" telescope someone received for Christmas. They looked good at first glance and seemed wonderful to the inexperienced, but they inevitably brought great disappointment. Brick-and-mortar stores are fading away in this digital era, and there are no longer department stores in many places. Now, a "department store" telescope represents any inexpensive, poorly made telescope that is difficult to use, with a wobbly mount and blurred images. These are often small refracting telescopes with a one-and-a-half- to two-inch objective lens on one end of the tube and an eyepiece for viewing at the other end.

In previous times, telescope magnification was greatly hyped. "See the skies with five hundred power or even a thousand power . . ." Unless the optics are superb and the telescope objective lens or its mirror are quite large by amateur standards, twelve inches or greater in diameter, a higher power beyond approximately 150× is pointless. The image is too dim and too poorly focused to be useful. Often, it is also further degraded by atmospheric turbulence ("poor seeing") and a wobbly mount. These days, power is less hyped. Reputable manufacturers do not stress power in their advertising, and they design their instruments to use powers best suited for their optical systems.

Today, even small telescopes from "department stores" often have good optics and maybe even a decent mount. Despite this, someone unfamiliar with using a telescope must overcome a steep learning curve. Telescopes are great but look long and hard before you leap.

Whether you get a telescope or stick to using your eyes and a planisphere, there is one unfortunate factor that will affect your view of the night sky: light pollution. The reddish-yellow sky glow hanging over large cities and suburbs, light pollution greatly diminishes the grandeur of the night sky. Excessive nighttime lighting, combined with poor design and placement of many outdoor lights, dims all but the brightest stars. It is a significant reason why amateur astronomy is not thriving as we would wish. People growing up today live in mostly urban areas and never get a chance to see a star-filled sky.

It doesn't have to be like this. Consider learning more about light pollution and join those seeking to fight it and improve the quality of nighttime lighting. Check out the IDA and similar environmental groups. It will take a generation, but we can improve the lighting in cities enough to restore many of the lost stars.

Acknowledgments

DAVID LEVY, the renowned comet discoverer and author, suggested my name for the Sky Spy columns in the *Caliente* section of the *Arizona Daily Star*. David has been a wonderful astronomy buddy and good friend for many years. Inger Sandal, the former editor of *Caliente*, allowed me to write my columns and always kindly overlooked my foibles and missteps along the way.

For fear of forgetting someone, I will not try to list the many astronomy friends who helped me along the way, except for James McGaha, a retired pilot and superb astronomer, who has suffered the pains and joys of amateur astronomy and observatory operation with me for years. Alexis Huicochea, *Arizona Daily Star* Business, Metro, and Features team leader, and Gerald Gay, assistant editor of *Caliente*, kindly have continued to support the Sky Spy columns in *Caliente* and have most graciously allowed me to use material from past columns for this book.

Heather Jacobson provided superb copyediting for the manuscript. Nevertheless, I alone am responsible for all of the mistakes that have slipped through the editing process for the book despite Heather's work and the incredible support I have received from Allyson Carter, senior acquisitions editor, and Amanda Krause, editorial, design, and production manager for the University of Arizona Press.

Astronomical Resources

T HERE ARE countless books, magazines, mobile apps, and internet resources for those interested in any hobby or undertaking. Enjoying a star-filled sky is a natural human trait. Learning more about the sky is fun and comes easily to some. It takes a bit of effort for others to learn the constellations, be able to find planets, and understand routine astronomical phenomena. That doesn't matter. Do as much or as little as you wish. Spend as much time and money or as little time and money as you wish. It is a hobby and should be enjoyed. However, to get really knowledgeable takes a bit of work and research. Most people find this enjoyable if they have the proper resources or are pointed in the right direction.

This brief section lists those astronomical resources I have found to be extremely helpful, and I refer to them all of the time. They are either inexpensive mobile applications or good astronomical internet sites.

Abrams Planetarium—Sky Calendar: https://www
.abramsplanetarium.org/
Amateur Astronomy magazine: https://amateurastronomy.com/
American Astronomical Society (AAS): https://aas.org/

Arizona Daily Star (home of *Caliente* and the Sky Spy columns):
   https://tucson.com

Astronomers Without Borders, One People, One Sky: https://www
   .astronomerswithoutborders.org/home

Astronomical League: https://www.astroleague.org/

Astronomy Picture of the Day: https://apod.nasa.gov/apod/astropix
   .html

Astronomy Technology Today: https://astronomytechnologytoday.com/

Cosmos—The SAO Encyclopedia of Astronomy: https://astronomy
   .swin.edu.au/cosmos/

Globe at Night: https://www.globeatnight.org/

Grasslands Observatory: https://www.3towers.com/

Heavens Above (the premier website for satellite information):
   https://www.heavens-above.com/main.aspx

International Dark-Sky Association (IDA): https://www.darksky
   .org/

Lowell Observatory (Flagstaff, Arizona): https://lowell.edu/

LunarPhase Pro: Moon phase prediction software. Many download
   versions for desktop, laptop, and mobile applications.

MOON Pro: a mobile application

Night Sky: a mobile application

Planetary Science Institute (PSI): https://www.psi.edu/

Royal Astronomical Society of Canada (RASC): https://rasc.ca/

The Sea and Sky—Astronomical Calendar: http://www.seasky.org/
   astronomy/astronomy-calendar-current.html

Sky and Telescope magazine: https://skyandtelescope.org/

SkySafari 7 Plus: a mobile application

Star Rover: a mobile application

Starry Night (astronomy charting and astronomy control software):
   https://www.starrynight.com/

Stars (Jim Kaler's wonderful site for information about hundreds of
   the brightest stars): http://stars.astro.illinois.edu/sow/sowlist.html

Tucson Amateur Astronomy Association, Inc. (TAAA):
   http://tucsonastronomy.org/

Glossary

THE FOLLOWING astronomical terms are in common use and often appear in the Sky Spy columns. Everyday terms, such as orbit, the Moon, the Sun, the solar system, and outer space are not included. It is assumed that the reader is quite familiar with them from daily experience. Likewise, more technical or obscure astronomical terms like "parsec" and "H II region" are not defined, as they are beyond the scope of this book, which is devoted to the simple enjoyment of the night sky coupled with a basic understanding of what is being observed. Like any glossary, the list is selective and somewhat arbitrary; this is not a dictionary. Hopefully, the list will be inclusive enough for the readers of this book.

ALTITUDE. Altitude and azimuth are part of the horizontal coordinate system. Altitude is a measure of the angular distance of an object above or below the horizon. Objects located on the horizon have 0 degrees altitude, and at the zenith (directly overhead) 90 degrees altitude. In formal astronomical circles, objects below the horizon have negative altitudes.

ANNULAR SOLAR ECLIPSE. A solar eclipse in which the Moon passes directly in front of the Sun but does not completely cover the Sun's entire disk. This leaves a "ring of fire" or annulus around the Moon.

APHELION. The point at which Earth or another body is farthest from the Sun on its orbit around the Sun. Earth's aphelion occurs in early July.

APOGEE. The point at which a body orbiting Earth is farthest from Earth. This is usually applied to the Moon or an artificial satellite of Earth. Apogee may also describe the point at which a body is farthest from another body about which it orbits.

ASTERISM. A pattern of stars, such as a triangle or rectangle of stars, that is not a formally recognized constellation.

ASTEROID. A small rocky body orbiting the Sun. Known asteroids number in the hundreds of thousands, with many smaller ones yet to be discovered. They range in size from over three hundred miles (five hundred kilometers; the diameter of Vesta) to only a few meters in diameter. Most asteroids lie in the asteroid belt between Mars and Jupiter.

AURORA. An electrical glow in a planet's atmosphere caused by the interaction of the planet's magnetic field with charged particles from the Sun. Regarding Earth, auroras are most common in high northern and southern latitudes, where Earth's magnetic field is more concentrated. In the Northern Hemisphere, they are named aurora borealis or the northern lights. In the Southern Hemisphere, they are named aurora australis or the southern lights. They may be quite bright and display a multitude of quickly changing forms, streamers, arcs, curtains, and rays, with a variety of colors, most often green and red.

AZIMUTH. See altitude.

BACKGROUND (FIXED) STARS. Except for the Sun, all other stars are so far away they appear fixed in place and do not move compared to "foreground" objects such as the Moon and planets, which move with respect to the background stars.

BLUE MOON. A colloquial term for a rare event. Astronomically, a blue Moon is a third full Moon in a season with four full Moons, when ordinarily there are only three full Moons in a season. A blue Moon is also defined as a second full Moon in a calendar month. Blue Moons are uncommon but not particularly rare.

CELESTIAL COORDINATE SYSTEM. An equatorial coordinate system to define the position of an object in the sky. Right ascension is the equivalent of longitude, and declination is the equivalent of latitude.

CELESTIAL EQUATOR. The projection of Earth's equator into space, dividing the sky into the north celestial hemisphere and the south celestial hemisphere.

CELESTIAL POLES. The projection of the North and South Poles of Earth into the sky. Polaris is close to the north celestial pole and is called the "pole star" in the Northern Hemisphere. The sky appears to rotate about the celestial poles just as Earth rotates around its poles.

CIRCUMPOLAR. Refers to stars and constellations near enough to the north or south celestial pole for one's latitude that they never dip below the horizon as Earth rotates. They never rise or set.

CLASSICAL PLANET. As opposed to a dwarf planet. Mercury, Venus, Earth, Mars, Jupiter, Saturn, Uranus, and Neptune are the classical planets. Pluto was demoted from being a classical planet to a dwarf planet.

COMET. Comets are large "dirty snowballs" consisting of various ices of frozen water, carbon monoxide, carbon dioxide, ammonia, and methane mixed with dust and rocky debris. This icy dusty mixture is the "nucleus" of the comet that is stable in the cold outer reaches of the solar system. When a comet travels into the inner portion of the solar system, at or inside of Earth's orbit, the nucleus is heated by the Sun and gives off large volumes of thin gas as its ices vaporize.

CONJUNCTION. A situation in which two astronomical objects are close to each other in the sky.

CONSTELLATION. A specific, identifiable grouping of stars, traditionally representing a mythical person or creature. The modern definition of constellations with demarcation of their borders was established by the International Astronomical Union in 1930.

COORDINATED UNIVERSAL TIME (UTC). The official time standard for the world. It is roughly equivalent to Greenwich mean time (GMT), the time at 0 longitude at Greenwich, England. It uses a twenty-four-hour clock notation and has no daylight saving.

CRESCENT. The curved, sickle shape of the Moon or other planetary body when it is less than half lit.

DECLINATION. The celestial equivalent of terrestrial latitude in the celestial coordinate system.

DEEP SKY OBJECT (DSO). A classification used in amateur astronomy for any celestial object that is not an individual star or a member of the solar system. The term usually refers to faint galaxies, nebulae, and star clusters.

DOUBLE STAR. A pair of stars close to each other from our viewpoint. They may be revolving around a common center of gravity and are technically a binary star system, or they may be unrelated to each other and simply lie along the same line of sight from Earth.

DWARF PLANET. As opposed to a classical or major planet. A controversial term used by the International Astronomical Union to classify an object orbiting the Sun that has enough mass to form a somewhat spherical body but not enough to "dominate" its region of space. Pluto with a diameter of 1,476 miles (2,376 kilometers) and Ceres with a diameter of 592 miles (952 kilometers) are the most well-known dwarf planets.

EARTHSHINE. A glow on the unlit part of the Moon. It is best seen a few days before or a few days after new Moon. Leonardo da Vinci was the first person to realize earthshine is light from the Sun reflected off of Earth onto the Moon and back to Earth.

ECLIPTIC. The plane of Earth's orbit around the Sun. It is the path in the sky the Sun follows on its apparent yearly movement. Most of the planets and major asteroids lie within a few degrees of the ecliptic.

ELONGATION. Distance east or west of the Sun as viewed from Earth, usually in degrees. This term is commonly used to describe the distance of Mercury or Venus from the Sun in the morning or evening sky.

EQUINOX. The instant in time when the Sun appears to cross the equator from south to north (spring or vernal equinox around March 21) or from north to south (fall or autumnal equinox around September 21). The day upon which the equinox occurs is often referred to as the equinox, though the equinox is technically a brief moment in time.

EVENING STAR. A bright planet, usually Venus, that is prominent in the twilight evening sky. Equivalently, a morning star is a planet bright in the twilight morning sky.

FALLING STAR. See meteor.

FIXED STARS. See background (fixed) stars.

FULL MOON. The Moon is fully illuminated by the Sun. It sits on the opposite side of Earth from the Sun, 180 degrees from the Sun.

GALAXY. A large collection of millions or billions of stars and dust and gas. When the term is capitalized (Galaxy), it refers to the Milky Way, our parent galaxy.

GIBBOUS. Gibbous refers to the Moon or other planetary body that is more than half lit but not completely lit like the full Moon.

GREENWICH MEAN TIME (GMT). See coordinated universal time.

HARVEST MOON. The full Moon nearest to the autumnal (September) equinox.

HUNTER'S MOON. The first full Moon after the Harvest Moon.

INFERIOR PLANET. An archaic term for Mercury or Venus, which orbit closer to the Sun than Earth.

INTERNATIONAL ASTRONOMICAL UNION (IAU). The internationally recognized body for professional astronomers and national astronomical societies.

KELVIN. The standard temperature scale used in science for extremely low or extremely high temperatures. Absolute zero, the coldest known temperature, is 0 Kelvin or –273.16 degrees Celsius. The freezing point of water is 0 degrees Celsius or 273 Kelvin.

KITT PEAK NATIONAL OBSERVATORY (KPNO). A U.S. national observatory located on Kitt Peak of the Quinlan Mountains in the Arizona Sonoran Desert. It is on the Tohono O'odham Nation fifty-five miles (eighty-eight kilometers) west of Tucson, Arizona. It is one the largest astronomical facilities in the Northern Hemisphere.

KUIPER BELT (ALSO KNOWN AS EDGEWORTH-KUIPER BELT). A vast region in the cold outer reaches of the solar system beyond Neptune where millions of small icy bodies with varying amounts of rock orbit the Sun. Most comets with long periods (greater than roughly two hundred years) probably originate in the Kuiper belt.

LIGHT POLLUTION. A pervasive glow in the night sky that dims faint stars and celestial objects. It is caused by artificial nighttime lighting and is a significant problem in most urban and suburban regions of the world.

LIGHT TRESPASS. Light shining where it is not wanted, such as a neighbor's floodlight shining in your bedroom window.

LIGHT YEAR. The distance light traveling through a vacuum covers in a year, approximately 5.88 trillion miles or 9.46 trillion kilometers.

LUMINOSITY. The amount of energy emitted by a star.

LUNAR ECLIPSE. A lunar eclipse occurs when the Moon passes into Earth's shadow. A total lunar eclipse takes place when the Moon fully enters the umbra of Earth's shadow. If the Moon does not fully pass through Earth's umbra, there is a partial lunar eclipse. Lunar eclipses take place at full Moon.

LUNAR MONTH. The time between successive new Moons or successive full Moons, approximately 29½ days.

MAGNITUDE. The astronomical term for star brightness. The modern system of magnitudes has five magnitude steps that correspond to an exact difference in brightness of a factor of one hundred. Each magnitude is 2.512 times dimmer than the preceding magnitude. Very bright objects, such as the Sun (−26.8), Jupiter (−2.64), and Sirius (−1.46) have negative magnitudes. The first-magnitude stars range from Sirius to Regulus (magnitude 1.35). The second-magnitude stars range from 1.51 to 2.5 magnitude, and so forth.

MAJOR PLANET. See classical planet.

MERIDIAN. An imaginary line that runs from the north to the south celestial pole through the observer's zenith.

MESSIER OBJECT. French comet hunter Charles Messier (1730–1817) compiled a list of 103 bright celestial objects that at first look resembled comets but in fact were fixed celestial objects. Most of the brightest star clusters, nebulae, and galaxies are Messier objects and are listed as M1 to M103. Another seven were added to the list in modern times, as they were described by Messier in his writings but did not appear on his published list.

METEOR. Falling star or shooting star. The visible passage of a small piece of rock or dust from outer space as it heats up and vaporizes while traveling through the atmosphere.

METEOR SHOWER. A series of meteors that radiate from a localized area of the sky. A meteor shower is named after the constellation in which its radiant is located. Famous meteor showers include the Perseids in August, the Leonids in November, and the Geminids in December.

METEORITE. A meteor that has reached the ground.

MIDNIGHT. Astronomically speaking, midnight is the opposite of solar noon.

MILKY WAY. Our parent galaxy. When we look toward the Milky Way, we see our parent galaxy on edge, viewing thousands of stars seemingly piled on top of one another, thus giving the illusion of milk spilled (*via lactea* in Latin) across the sky.

MINOR PLANET. Another name for an asteroid.

MORNING STAR. See evening star.

NADIR. A low point. Astronomically speaking, it represents the point directly opposite zenith.

NAVIGATOR'S TRIANGLE. See Summer Triangle.

NEBULAE. Interstellar clouds of hydrogen, helium, other mostly ionized gases, and dust particles.

NEW MOON. When the Moon has no visible illumination from the Sun. The Moon is directly between the Sun and Earth.

NOON. In an astronomical sense, noon refers to solar noon when the Sun crosses the meridian and reaches its highest point in the sky.

OCCULTATION. The covering of a distant astronomical body by a closer body, such as the Moon occulting the star Regulus. Lunar and solar eclipses are special types of occultations.

OORT CLOUD (ALSO CALLED ÖPIK-OORT CLOUD). The most distant region of the solar system, a vast sphere that contains trillions of small icy or rocky bodies.

OPPOSITION. The position of a planet when it is exactly opposite the Sun in the sky as seen from Earth. A planet at opposition is at its best for observing from Earth.

PENUMBRA. The light outer part of a shadow where the light from a celestial body, such as the Sun or Moon, is not completely blocked, as opposed to the umbra where there is a complete blocking of the light. A typical celestial shadow consists of an umbra surrounded by a penumbra.

PERIGEE. The point at which a body orbiting Earth is closest to Earth. This is usually applied to the Moon or an artificial satellite of Earth. Perigee may also describe the point at which a body is closest to another body about which it orbits.

PERIHELION. The point at which Earth or other body is closest to the Sun on its orbit around the Sun. Earth's perihelion occurs in early January.

PHASE OF THE MOON. The shape of the sunlit portion of the Moon as seen from Earth.

PLANET. For the purposes of this book, planet refers to Mercury, Venus, Earth, Mars, Jupiter, Saturn, Uranus, and Neptune, the classical planets, which are large celestial bodies that orbit around the Sun. Pluto at one time was considered a (classical) planet but is now known as a dwarf planet. Planets orbit many other stars, but those outside the solar system are not discussed in this book.

PLANISPHERE. A plastic or cardboard chart showing the constellations.

RIGHT ASCENSION. The equivalent of longitude in the celestial coordinate system.

SEEING. An astronomical term describing the steadiness of the atmosphere. This depends on many factors including one's location and the local weather. Good seeing represents a steady atmosphere that enables easier telescopic viewing or imaging of small details. When the seeing is especially poor, even the best telescope may show the Moon and planets as nothing more than blurry blobs. When the seeing is exceptionally good, even a mediocre telescope may show especially good detail on the Moon and planets.

SHOOTING STAR. See meteor.

SOLAR ECLIPSE. A solar eclipse occurs when the Moon passes in front of the Sun. A partial solar eclipse occurs when the Moon only partially covers the Sun. A total solar eclipse occurs when the Moon completely covers the Sun. It is one of the most spectacular events in nature. Solar eclipses happen at new Moon.

SOLSTICE. The solstices are those points at which the Sun is at its greatest distance north (summer solstice) or south (winter solstice). Solstice means a culminating point or turning point. The solstices take place at a brief specific time, but the day upon which a solstice occurs is often also referred to as the solstice or solstice day. The summer solstice takes place around June 21 every year, and the winter solstice takes place around December 21. The solstices are reversed in the Southern Hemisphere, making the seasons opposite those of the Northern Hemisphere.

STAR. A large collection of gas that has enough mass to contract upon itself, generating intense internal temperatures through nuclear reactions that heat up the gas further until it emits large amounts of radiation, mostly in the visible portion of the electromagnetic spectrum. The Sun is the nearest star. Stars are distant suns.

STAR CLUSTER. A collection of stars that are usually gravitationally bound to each other and traveling through space together. The Pleiades are an example of a bright loose collection of stars known as an open cluster. Open clusters contain a few hundred to a few thousand stars. Globular clusters are very compact collections of hundreds of thousands of stars.

SUMMER TRIANGLE (NAVIGATOR'S TRIANGLE). An asterism of three bright stars, Altair, Vega, and Deneb.

SUNRISE. The moment the upper edge of the Sun's disk becomes visible above the horizon.

SUNSET. The moment the upper edge of the Sun's disk disappears below the horizon.

SUPERIOR PLANET. An archaic term for those planets (Mars, Jupiter, Saturn, Uranus, and Neptune) that orbit farther from the Sun than Earth.

SUPERMOON. When a full Moon occurs around perigee, the point at which the Moon is closest to Earth. A supermoon may appear to be slightly larger and slightly brighter than a normal full Moon.

SUPERNOVA. A giant "explosion" that occurs at the end of life for stars that are several times more massive than the Sun. This stellar explosion can emit as much light as the rest of the stars in its parent galaxy combined, for some days. The dying star finally collapses into a small dense star composed of neutrons, a neutron star, or into a black hole, a tiny space so dense and massive that not even light can escape from it.

TERMINATOR. The boundary between the lit (day) side of a celestial body, such as the Moon, and its unlit (night) side.

TRANSIT. Passage of a celestial body in front of another celestial body, such as a transit of Mercury, where Mercury appears as a tiny black dot slowly passing in front of the Sun.

TROPICS. The term "tropic" is derived from a Greek word meaning turn, change of direction, or change of circumstance, because the Sun appears to turn back at the solstices and reverse its direction in the sky. The Tropic of Cancer (also known as the Northern Tropic, at 23.44 degrees north latitude) is the north circle of latitude at which the Sun appears directly overhead at its most northern point at the summer solstice. The equivalent line of latitude south of the Equator is called the Tropic of Capricorn at 23.44 degrees south at the winter solstice. The region between these two lines is the "tropics."

TWILIGHT. The time between day and night. It is caused by scattered sunlight in the atmosphere, which illuminates the sky and the ground when the Sun is not far below the horizon. Civil twilight is the time when the center of the Sun's disk is less than 6 degrees below the horizon. During this time, the brightest stars and planets appear. At nautical twilight the Sun is between 6 and 12 degrees below the horizon. Astronomical twilight takes place when the center of the Sun is between 12 degrees and 18 degrees below the horizon. The sky appears nearly fully dark, but very faint objects may still be hidden by the dim sunlight in the atmosphere.

WAXING AND WANING. Referring to the lit portion of the Moon. When the Moon's lit portion is increasing, it is said to be waxing, and when the Moon's lit portion is decreasing, it is said to be waning. These terms can also be generically used for other celestial bodies.

ZENITH. Directly overhead. See also altitude.

ZODIAC. A strip of sky that stretches about 9 degrees on either side of the ecliptic, in which the Moon and planets can generally be found.

ZODIACAL LIGHT. A large, faint cone of light visible in the east before sunrise or in the west after sunset. It is caused by sunlight reflecting off of tiny dust grains and ice particles in the plane of the solar system.

Selected Bibliography

Jim Kaler's Stars website: http://stars.astro.illinois.edu/sow/sowlist.html

The Sea and Sky's Astronomy Reference Guide: http://www.seasky.org/astronomy/astronomy-glossary.html

Sky and Telescope magazine's Astronomy Terms: https://skyandtelescope.org/astronomy-terms/

Swinburne University's Cosmos—The SAO Encyclopedia of Astronomy: https://astronomy.swin.edu.au/cosmos/

timeandate.com's Astronomical Glossary: https://www.timeanddate.com/astronomy/explanation-terms.html

Wikipedia's Glossary of Astronomy: https://en.wikipedia.org/wiki/Glossary_of_astronomy

Index

Note: Page numbers followed by *f* indicate a figure on the corresponding page.

Achernar, 60, 96–97
Albireo, 71, 72*f*
Alcor, 77, 91
Aldebaran, 61*f*, 62, 108–9
Aldrin, Buzz, 133
Algol, 63, 103
Alhena, 63
Almach, 102
Alnilam, 68, 69*f*
Alnitak, 68, 69*f*
Alpha Centauri, 52, 100
Alpha Pavonis (Peacock Star), 78–79
Alphard, 63, 98
Alpheratz, 102
Altair, 82, 86, 100, 148
altitude, 8–9, 12, 154
American Astronomical Society
    (AAS), 138–39, 140
Andromeda the Maiden, 72*f*, 88*f*, 89,
    93, 102–3
angular distance, 7

annular eclipses, 53
Antares, 66, 74
apogee, 8, 27
Apollo 11 Moon Landing, 133–34, 145
Aquarius the Water Carrier, 74*f*, 86
Aquila the Eagle, 86, 87*f*, 113
Arcturus, 67, 91
Aries the Ram, 88*f*, 89, 105
Armstrong, Neil, 133
asterism, 82
asteroids (minor planets), 46
Astronomical League, 164
Auriga the Charioteer, 61*f*, 62*f*, 89–90,
    99, 113
azimuth, 8–9, 12

Bastille Day, 132–33
Beehive, 6, 65*f*, 92, 116
Bellatrix, 68, 69*f*
Belt of Venus, 110
Beta Centauri, 100

Betelgeuse, 68–69, 89

Big Dipper (Ursa Major the Great Bear), 90–91, 99, 101

binoculars: buying, 163; optical aid, 109; tripod, 6, 47; viewing the Andromeda Galaxy, 103; viewing asteroids, 46; viewing the Beehive (M44), 92; viewing Coma Berenices, 94; viewing the Double Cluster, 104; viewing Gacrux, 76; viewing Jupiter's moons, 47; viewing the Kuiper belt and the Oort cloud, 50; viewing Mars, 45; viewing the Messier objects, 143; viewing the Milky Way, 115; viewing Mizar and Alcor, 77; viewing the Moon, 17, 23, 28, 30; viewing Omega Centauri, 100; viewing Ophiuchus the Serpent Holder, 101; viewing the "Peacock Star" Alpha Pavonis, 78; viewing the Pleiades, 106; viewing satellites, 155; viewing Taurus the Bull, 62, 108; viewing Uranus, 37, 48; viewing Venus, 44

blood Moon, 32

blue Moon, 26

Boötes the Herdsman, 91

calendar systems, 14, 19–20, 24, 126–28

Cancer the Crab, 92

Canis Major the Greater Dog and Canis Minor the Lesser Dog, 93, 98, 99, 100, 113

Canopus, 80, 99

Capella, 70, 89

Capricornus the Sea Goat, 74*f*, 86

Cassiopeia the Queen of Ethiopia, 88*f*, 93–94, 113

Castor, 70–71

celestial coordinate system, 9

celestial equator, 9

Centaurus the Centaur, 100

Cepheus the King of Ethiopia, 93–94, 113

Ceres, 46

Cetus the Whale, 94

Christmas, 129

circumpolar constellation, 91

Collins, Michael, 133

Columba the Dove, 99

Coma Berenices, 94

Comet Halley, 118

comets, 116–21, 142, 143

conjunctions, 38, 41*f*

constellations: Andromeda the Maiden, 72*f*, 88*f*, 89, 93, 102–3; Aquarius the Water Carrier, 74*f*, 86; Aquila the Eagle, 86, 87*f*, 113; Aries the Ram, 88*f*, 89, 105; Auriga the Charioteer, 61*f*, 62*f*, 89–90, 99, 113; Big Dipper (Ursa Major the Great Bear), 90–91, 99, 101; Boötes the Herdsman, 91; brightness order of stars, 57; Cancer the Crab, 92; Canis Major the Greater Dog and Canis Minor the Lesser Dog, 93, 98, 99, 100, 113; Capricornus the Sea Goat, 74*f*, 86; Cassiopeia the Queen of Ethiopia, 88*f*, 93–94, 113; Centaurus the Centaur, 100; Cepheus the King of Ethiopia, 93–94, 113; Cetus the Whale, 94; circumpolar, 91; Columba the Dove, 99; Coma Berenices, 94; Corona Australis the Southern Crown, 92; Corona Borealis the Northern Crown, 91–92, 97; Corvus the Crow, 95, 98; Crater the Cup, 98; Crux the Southern Cross, 76; Cygnus the Swan, 95–96, 113;

definition, 9, 84; Double Cluster, 104, 116; Draco the Dragon, 96; Eridanus the River, 96–97; Gemini the Twins, 62*f*, 97, 113; Hercules the Hero, 97; Hydra the Water Snake, 98; Leo the Lion, 98; Lepus the Hare, 99; Little Dipper, 90–91, 96; Lynx the Lynx, 99; Lyra the Lyre, 97, 99–100; Monoceros the Unicorn, 100; Ophiuchus the Serpent Holder, 101, 113; Orion the Hunter, 62*f*, 89–90, 99, 100, 101–2, 113; Pavo the Peacock, 78; Pegasus the Winged Horse, 72*f*, 88*f*, 102, 105; Perseus the Hero, 93, 103, 104, 113; Pisces the Fishes, 88*f*, 105; Piscis Austrinus the Southern Fish, 74*f*; planisphere, 5, 85; the Pleiades, 105–6, 108, 116; Sagittarius the Archer, 101, 106–7, 113; Scorpius the Scorpion, 101, 108; seasons, 85; Sextans the Sextant, 98; stars in, 84; Summer Triangle, 72*f*, 100, 148–49*f*; Taurus the Bull, 61*f*, 62*f*, 108–9; Triangulum the Triangle, 88*f*, 89; Ursa Major the Great Bear (Big Dipper), 90–91, 99, 101; Virgo the Virgin, 98, 109; zodiacal constellations, 13
coordinated universal time, 128
Corona Australis the Southern Crown, 92
Corona Borealis the Northern Crown, 91–92, 97
Corvus the Crow, 95, 98
Crater the Cup, 98
crescent, 10
Crux the Southern Cross, 76
Cygnus the Swan, 95–96, 113

D-Day, 131–32
declination, 9
Deneb, 57, 71–72*f*, 82, 96, 100, 148
Deneb Kaitos, 73
Denebola, 57, 65*f*, 73, 98
dog days of summer, 134
Double Cluster, 104, 116
Draco the Dragon, 96
Dschubba, 73–74, 108
dwarf planet, 36

Earth: earthshine, 29–30; Earth's shadow, 110, 111*f*; and Mars, 45; and meteors, 116, 119; Moon orbiting, 8, 10, 16–17, 19, 23; Moon stabilizing Earth's axis, 14–15; and Neptune, 49; orbiting Sun, 8, 10–11, 16, 24, 123, 125; and Polaris, 77; and Saturn, 48; and solar eclipses, 53–55; solar energy, 52; umbra, 32, 34
earthshine, 29–30
Earth's shadow, 110, 111*f*
Easter, 129–30
ecliptic path, 10
Eddington, Arthur, 139–40
elliptical orbit, 10
elongation, 10, 38–39*f*, 154
equinox, 10–11, 123–24, 125
Eridanus the River, 96–97
Eris, 46
evening star, 37–38

falling star, 116
Fomalhaut, 74, 75*f*
full Moon: cycle, 20–21, 127; desirable for D-Day, 131–32; and Easter, 129–30; and Mars size, 152; Metonic cycle, 24; names, 24–25; and Passover, 129–30; photo of sunset, 111*f*

Gacrux, 74f, 75f, 76
Gagarin, Yuri, 139
Geminids, 121–22
Gemini the Twins, 62f, 97, 113
gibbous phase, 11
Globe at Night, 136–37
Gomeisa, 79, 93
Graffias, 73, 108
Great Orion Nebula, 102, 116

Harvest Moon, 25–26
Haumea, 46
Hercules the Hero, 97
Herschel, William, 138, 141
Hoffleit, Ellen Dorrit, 140
Hubble Space Telescope, 70, 150, 162
Hunter's Moon, 25–26
Hyades, 62, 108–9, 116, 139
Hydra the Water Snake, 98

Ides of March, 130
Independence Day, 129, 132
International Astronomical Union
    (IAU), 9, 13, 36, 56, 140–41, 150
International Astronomy Day, 136
International Dark-Sky Association
    (IDA), 141
International Space Station (ISS), 144,
    150, 152, 155–56

Jupiter, 37–38, 41f, 45, 46–48, 118, 152

Kelvin scale, 51, 53, 58–59, 62, 66–67, 82
Kochab, 76
Kuiper belt, 46, 49–50

Labor Day, 129
Leonids, 120–21
Leo the Lion, 98
Lepus the Hare, 99

light years, 35, 52, 62–63, 68–70, 73–74,
    79–80
Little Dipper, 90–91, 96
Lowell, Percival, 142
Lowell Observatory, 49, 142, 146
lunar eclipses, 3, 32–34
lunar month, 16–17, 19–20, 24–25, 127
Lynx the Lynx, 99
Lyra the Lyre, 97, 99–100

magnitude, 55–56
Makemake, 46
Mars, 37–38, 45, 46, 66, 118, 152
Martin Luther King Day, 129
Memorial Day, 129, 130–31
Mercury, 35, 38–39f, 40–41f, 42, 44, 135
meridian, 11
Messier, Charles, 142–43, 150
Messier objects, 143, 150
meteorites, 119
meteors, 116, 119–22
Metonic cycle, 24
Milky Way: and the Andromeda Gal-
    axy, 103; and Aquila the Eagle, 86;
    and Arcturus, 67; black hole, 112;
    and Cassiopeia the Queen of Ethi-
    opia, 94; and Cygnus the Swan, 96;
    Galileo's observations of, 47; "Great
    Rift," 86, 87f; and Omega Centauri,
    100; and Ophinuchus the Serpent
    Holder, 101; and Orion the Hunter,
    102; and Sagittarius the Archer,
    107; and Scorpius the Scorpion,
    108; Sky Spy columns, 153; stars in,
    59, 82, 113; summer, 112–13, 114f, 115,
    133, 161; winter, 90, 93, 112–13, 115
Mintaka, 68, 69f
Mirach, 102
Mira (Omicron Ceti), 77, 94
Mirfak, 103

Mitchell, Maria, 143
Mizar, 77, 91
mobile apps, 5–6
Monoceros the Unicorn, 100
Moon: Apollo 11 Moon landing, 133–34, 145; blood Moon, 32; blue Moon, 26; calendar systems, 14, 19–20, 24, 127; conjunctions, 38, 41f; Copernicus, 28; craters, 28; crescent Moon, 10, 15, 16, 19–21, 29–30, 41f; earthshine, 29–30; eclipses, 3, 32–34; gibbous Moon, 11, 16, 20–21, 29, 44f, 131; guide to celestial objects, 3; halos, 30–32; Harvest Moon, 25–26; Hunter's Moon, 25–26; lunar eclipses, 3, 32–34; lunar libration, 17; lunar month, 16, 20; lunar occultations, 23–24, 62, 82; maria, 28; and Mercury, 42; Metonic cycle, 24; micromoon, 27; Moon illusion, 23; moonrise, 21–23; moonset, 21–23; new Moon, 16, 19; orbit, 8, 10, 16; penumbra, 32; phases, 17–18, 21, 127, 147, 154, 155; and Pluto, 49; quarter Moon, 19f, 20–21, 30, 133, 135, 152; reflected sunlight, 15; religious importance, 129–30; Sky Spy columns, 15; and solar eclipses, 53–55; stabilizing Earth's axis, 14–15; and Sun, 15, 21–23; supermoon, 26–27; terminator, 30; Tycho, 28; and Venus, 44; waxing and waning, 12. *See also* full Moon; quarter Moon
morning star, 37–38

National Aeronautics and Space Administration (NASA), 143–44, 150
Navigator's Triangle (Summer Triangle), 83

nebulae, 68
Neptune, 37, 48–49
new Moon, 16, 19
Nightingale, Mike, 133
Northern Hemisphere: astronomical observation, 3, 85, 113; cardinal sky directions, 18; equinox, 10; Harvest Moon, 25; leans toward or away from Sun, 123, 124f; north celestial pole, 77, 90; sky, 99; Sky Spy columns, 85; soltices, 11
North Pole, 77, 123

Omega Centauri, 75f, 100, 116
Omicron Ceti (Mira), 77, 94
Oort cloud, 50
Ophiuchus the Serpent Holder, 101, 113
opposition, 40–42
Orion the Hunter, 62f, 89–90, 99, 100, 101–2

partial solar eclipses, 55
Passover, 129–30
Pavo the Peacock, 78
Payne-Gaposchkin, Cecilia, 144–45
Peacock Star (Alpha Pavonis), 78–79
Pegasus the Winged Horse, 72f, 88f, 102, 105
perigee, 8, 26–27
Perseids, 119–20
Perseus the Hero, 93, 103, 104, 113
Pisces the Fishes, 88f, 105
Piscis Austrinus the Southern Fish, 74f
Pi Scorpii, 73, 108
Planetary Science Institute (PSI), 145
planets: appearance, 36; appearance at twilight, 12; asteroids (minor planets), 46; axes, 15; conjunction, 38, 40, 41f; crossing the meridian,

11; definition, 35; distance, 35; dwarf planet, 36; ecliptic path, 10; elongation, 38, 39f; evening star and morning star, 37–38; Jupiter, 37–38, 41f, 45, 46–48, 118, 152; lit by reflected sunlight, 15; Mars, 37–38, 45, 46, 66, 118, 152; Mercury, 35, 38–39f, 40–41f, 42, 44, 135; Neptune, 37, 48–49; observing using mobile apps, 5; occultations, 23; in opposition, 40–42; Pluto, 36, 46, 49, 50, 142, 146; Saturn, 41f, 47–48, 49, 161; transiting the sun, 40; traveling through the zodiac, 13; Uranus, 37, 48, 141–42; Venus, 35, 37–38, 40–41f, 44–45, 135, 152

planisphere, 5–6, 85, 134, 162, 165

Pleiades, 105–6, 108, 116

Pluto, 36, 46, 49, 50, 142, 146

Polaris, 57, 77–78, 96

Pollux, 70–71

President's Day, 129

Procyon, 70, 79, 93, 98

Proxima Centauri, 52

Regulus, 65f, 79, 98

Rigel, 68–69, 89, 96, 99

right ascension, 9

Royal Astronomical Society of Canada (RASC), 164

Sagittarius the Archer, 101, 106–7, 113

Saiph, 68, 69f

Saturn, 41f, 47–48, 49, 161

Scorpius the Scorpion, 101, 108

Serpens Cauda, 113

Sextans the Sextant, 98

shooting star (meteor), 116

Sirius, 52, 57, 79–81f, 93, 129, 134

Sky Spy columns: *Arizona Daily Star*, 128–29, 148, 151, 153; astronomers interview, 157–60; astronomical anniversary date, 138; brightest stars, 60; coverage of telescopes, 152; email complaints, 153–54, 160; email questions, 151, 160; horizontal coordinate system, 9; learning experience, 147; Milky Way, 153; minimal mention of Uranus, Neptune, and Pluto, 48; the Moon, 15; Northern Hemisphere, 85; no telescope access, 6; proofreading, 148–50; satellite observations, 155–56; Thanksgiving, 135

sleep deprivation, 4

solar eclipses, 19, 34, 53–55, 139, 161

solstices, 11, 125–26

Spica, 75f, 82, 109

Sputnik, I, 145–46

star parties, 163–64

stars: Achernar, 60, 96–97; Albireo, 71, 72f; Alcor, 77, 91; Aldebaran, 61f, 62, 108–9; Algol, 63, 103; Alhena, 63; Almach, 102; Alnilam, 68, 69f; Alnitak, 68, 69f; Alpha, 57; Alpha Centauri, 52, 100; Alphard, 63, 98; Alpheratz, 102; Altair, 82, 86, 100; Antares, 66, 74; Arcturus, 67, 91; Bellatrix, 68, 69f; Beta, 57; Beta Centauri, 100; Betelgeuse, 68–69, 89; brightness, 51–52, 55–56; Canopus, 80, 99; Capella, 70, 89; Castor, 70–71; colors, 58–59; in constellations, 84; definition, 35; Deneb, 57, 71–72f, 82, 96, 100; Deneb Kaitos, 73; Denebola, 57, 65f, 73, 98; Dschubba, 73–74, 108; Eddington's work, 139–40; evening star, 37–38; falling star, 116; fixed,

35; Fomalhaut, 74, 75f, Gacrux, 74f, 75f, 76; Gomeisa, 79, 93; Graffias, 73, 108; Hoffleit, 140; hydrogen, 144; Kochab, 76; magnitude, 55–56; makeup, 51; meteors, 116, 119–22; Milky Way, 113–15; Mintaka, 68, 69f, Mirach, 102; Mirfak, 103; Mizar, 77, 91; morning star, 37–38; names, 56–58; Omicron Ceti (Mira), 77, 94; of Orion, 68, 101–2; Peacock Star (Alpha Pavonis), 78–79; Pi Scorpii, 73, 108; Polaris, 57, 77–78, 96; Pollux, 70–71; positions of, 139, 142; Procyon, 70, 79, 93, 98; Proxima Centauri, 52; Regulus, 65f, 79, 98; Rigel, 68–69, 89, 96, 99; Saiph, 68, 69f, shooting star (meteor), 116; Sirius, 52, 57, 79–81f, 93, 129, 134; Spica, 75f, 82, 109; star parties, 163–64; supernova, 52; temperature, 51; twinkling, 36, 152; Vega, 57, 82–83, 96, 97, 100; Zubenelgenubi, 83; Zubeneschamali, 83. See also Summer Triangle

Summer Triangle, 72f, 82–83, 100, 148–49f

Sun: and Achernar, 60; and Aldebaran, 62; and Alhena, 63; and Alphard, 63; and Antares, 66; and Arcturus, 67; and Betelgeuse, 68; calendar system, 126–27; and Canopus, 80; and Capella, 70; and comets, 117, 119; and Deneb, 71; and Deneb Kaitos, 73; and Denebola, 73; description, 52–53; and Dschubba, 74; Earth orbiting, 8, 10; earthshine, 29; and Earth's shadow, 110; Earth's tilt toward, 123; eclipses, 19, 34, 53–55, 139, 161; ecliptic path,

109; elongation, 10; equinoxes, 123–24, 156; and Fomalhaut, 74; and Gomeisa, 79; halos, 30, 31f; and Jupiter, 46–47; and Kochab, 76; lunar eclipse, 32–34; magnitude, 56; and Mercury, 42; and Moon, 15, 21–23; Moon's position relative to, 16; nearest star, 35; and Neptune, 49; opposition, 40–42; and the "Peacock Star" Alpha Pavonis, 79; and Perseus the Hero, 103; and Phaeton, 121; planets orbiting, 35–36; and Pluto, 49; and Polaris, 78; and Procyon, 79; reflected sunlight lighting the Moon, 15; and Regulus, 79; religious importance, 130; and Rigel, 69; and Saturn, 47–48; and Sirius, 80, 134; solar eclipses, 19, 34, 53–55, 139, 161; solstices, 125–26; soltices, 11, 157; and Spica, 82; and stars, 51–52; sunrise, 22, 38, 102, 110, 124, 134; temperature, 58–59; transit the Sun, 40; Tropic of Cancer and Tropic of Capricorn, 126; twilight, 12; and Venus, 44–45; and zodiacal light, 111–12

supermoon, 26–27
supernova, 52

Taurus the Bull, 61f, 62f, 108–9
telescopes: buying, 6–7, 152, 162–65; Hubble Space Telescope, 70, 150; large telescopes, 133, 142; objects appear upside down, 18; professional telescopes, 113; radio telescopes, 86, 107, 112; small telescopes, 142, 143, 161; times before telescopes, 118; viewing the Andromeda galaxy, 103; viewing asteroids, 46; viewing Coma Berenices, 94,

95*f;* viewing the Double Cluster, 104; viewing Jupiter, 47; viewing the Kuiper belt and Oort cloud, 50; viewing Leo the Lion, 98; viewing M13, 97; viewing Mars, 45; viewing Mercury, 42; viewing the Milky Way's black hole, 107, 112; viewing the Moon, 17, 23, 28, 30, 47; viewing Neptune, 37, 49; viewing Orion the Hunter, 102; viewing the Pleiades, 106; viewing Pluto, 36; viewing satellites, 155; viewing Saturn, 47–48; viewing stars, 35, 52, 56, 70, 71, 113; viewing Uranus, 37, 142; viewing Venus, 44; viewing Virgo the Virgin, 109; view of the Andromeda Galaxy, 88*f;* view of the Beehive, 92*f;* view of the Pleiades, 106*f*

Thanksgiving, 128, 135–36
Tombaugh, Clyde, 146
transiting the Sun, 40
Triangulum the Triangle, 88*f,* 89
tropics, 126
Tucson Amateur Astronomy Association, Inc. (TAAA), 136, 152, 163

twilight: definition, 12; evening, 4, 10, 37, 38, 110, 111, 145; length of, 12; morning, 111, 120, 145; predawn, 10, 37, 38, 110, 134

Uranus, 37, 48, 141–42
Ursa Major the Great Bear (Big Dipper), 90–91, 99, 101

Valentine's Day, 129
Vassar College Observatory, 143
Vega, 57, 82–83, 96, 97, 100, 148
Venus, 35, 37–38, 40–41*f,* 44–45, 135, 152
Vesta, 46
Veteran's Day, 129, 135
Virgo the Virgin, 98, 109

waxing and waning, 12

zenith, 12
zodiac, 13
zodiacal light, 111–12
Zubenelgenubi, 83
Zubeneschamali, 83

About the Author

Tim B. Hunter has been an amateur astronomer since 1950, and he is the owner of two observatories, the 3towers Observatory and the Grasslands Observatory (http://www.3towers.com). He is also a prime example of someone whose hobby has run amok, spending more time and money on it than common sense would dictate.

He has been the President of the Tucson Amateur Astronomy Association, Inc. (TAAA) and a member of the TAAA since 1975. He is also a past Chair of the Board of Trustees of the Planetary Science Institute (PSI). Since 1986, Tim Hunter has been interested in the growing problem of light pollution. In 1987, he and Dr. David Crawford founded the International Dark-Sky Association, Inc. (IDA). IDA is a nonprofit corporation devoted to promoting quality outdoor lighting and combatting the effects of light pollution. Since 2007, Tim has written a weekly Sky Spy column for the *Caliente* section of the *Arizona Daily Star*. Asteroid 6398 is named Timhunter.